# The Nature Of Humans

# The Nature Of Humans

## Why We Behave As We Do

Ron Newby

ISBN: 0692686614
ISBN 13: 9780692686614
Library of Congress Control Number: 2016906032
Ron Newby, Del Mar, CA

*For Mac*

# Table of Contents

# One

## Introduction

*Man is unique not because he does science, and he is unique not because he does art, but because science and art equally are expressions of his marvelous plasticity of mind.*

—Jacob Bronowski

It is well known that identical twins often have a similar nature, whereas fraternal twins may display contrary dispositions. Even identical twins raised apart from an early age are behaviorally very similar. Our nature is heavily influenced by cultural conditioning and environmental factors. Less appreciated are the other major components to our nature: our inherited genetic behavioral traits. All our genes are inherited directly from our parents and even from our ancient ancestors, genes that ensured survival of our species. Many of these ancient genes remain vibrant in humans today.

All behaviors are originate as genetically coded brain activity of a brain that has been evolving over millions of years. The neurological chemicals and their pathways of our brain controls the response level of our behaviors. Not all hu man brains are identical, exemplified by the diversity and levels of expression of our genetic behavioral traits. We have both agreeable and contentious dispositions. Some of us have excessive fear. Of great concern is that human behaviors are drastically impacting the planet and the future of humanity. There are genetic factors, of which we may not be fully aware, that affect our behavior. Would an evaluation of our nature provide insights as to why we do as we do?

What then is the nature of humans? Not only does everyone have some sort of an opinion, but there are countless books that cover this question. The word *nature* is derived from the Latin word *natura,* or the innate qualities, disposition and intrinsic characteristics of living organisms. Humans are certainly living organisms.

This book addresses two questions. Can our ancient paleo-history and the neurosciences provide some insight as to the underpinnings of our nature? Which behaviors and values may better serve humanity? This book is a personal and selective assessment of the nature of humans. It will not be all inclusive; that would be impossible and tedious. My aspiration for writing this book is to stimulate thought and discussion as to the nature of humans with consideration of humanities long-term survival.

It seems that I have always had curiosity about what life is and about human behaviors. Growing up I became aware of the atrocious behaviors of some humans as well as the compelling stories of individuals who have shown great generosity and compassion. What is it about humans that we are such a mixture of vices and virtues? It is very obvious that humans have severely altered our planet. No other species has been as disruptive as humans to the extent that our future is in doubt.

Unique to humans alone are our constructs: religion, politics, the sciences and the humanities. This book is the accumulation of my interest in science and other scholarly subjects, including art, and my deep concern over some human behaviors. The world's people are now facing an uncertain future. Our sustainability is in doubt. My great concern is that we may be approaching our "term limits". With thoughtful consideration I am proposing that for the betterment and continuation of humanity, consideration should be given to Progressive ideals.

My interest in biology probably commenced in my 8th grade science class (1950) at John Marshall Jr. High School in Pasadena, California. This class was, without doubt a pivotal class in my life. It was a general science class taught by Mr. Williams. It was pivotal because I was introduced to a new world. The Periodic Table was displayed with, at the time, about 100 known elements. Since then an additional 17 elements have been discovered. The emphasis of this class was the life sciences. I vividly recall time-lapse films of seedlings germinating. We examined microscopic organisms living in pond scum and fungi growing on petri dishes. The world of cooties transformed into an exciting world of life forms with unique characteristics. We were assigned to do projects. I decided to collect 25 native plants growing in the nearby Sierra Madre Mountains. I glued twigs with flowers and leaves onto a large cardboard sheet and labelled them with both the common name and the scientific name of each specimen.

In High School, Mr. Ball, my botany class teacher, furthered my interest in science. He had a large collection of plant tissues mounted on glass slides. I remember that the microscopes were much better than the ones from junior high. As I look back I believe that Mr. Ball's greatest asset was that he conducted thought-provoking class discussions. One topic I recall concerned the destructive practices of man. We were shown photographs of clear-cut timbering in our country and in Brazil. He also showed a photograph of the farmlands in California's San Joaquin Valley. He asked us: what was this valley like before man?

I have since asked: How would this planet appear if *Homo sapiens* never existed; if the lineage leading to humans ceased prior to the evolution of humans? A planet full of animals and plants but without the human species would appear quite different from the present state of this planet. Obviously there would be no man-made structures, no religions nor politics nor bickering and certainly no weapons of mass destruction. The planet would be pristine. Rivers would run free, lakes would be clear, forested regions would extend over vast areas, the coral reefs would be vibrant. The polar regions would have an abundance of ice and the skies would be clear. There would be a natural balance on this planet. But that is not the case. The truth is that we are here and we are the governors of this planet. An alien to this planet might be puzzled by the wide range of behaviors and wonder how such a species survives.

What is the nature of humans that we have, often intentionally, altered and polluted our habitat to the extent that it may become uninhabitable, or at the minimum unattractive? Consciously we know better, or at least we should know better. At times we are not conscious of our behaviors. We are not conscious of our heart beating. It goes on without our consideration. There are genetic behavioral traits that unconsciously dictate our behavior. If we suddenly are facing danger, we don't consciously dial up the fear response — a rush of adrenaline and either flee or flight. Fear just takes over at appropriate times.

Humans are a creative and resilient species capable of extraordinary achievements. Yet we have enslaved, tortured, waged war and now we are playing havoc with the environment. We have a diversity of behaviors. Some people display excessive anger, aggression and greed. Some people are compassionate and generous towards others. A few individuals are cunning and deceitful. We have an inordinate fear of death, unless we have a martyr fixation. We have a contentious nature with regard to politics and religion. The sciences and the humanities are manifestations of our greatness. We are certainly a mixture of numerous genetic traits. Our behavior is greatly influenced by what we have learned, our environment and the circumstances we may encounter. We know that nurturing greatly influences our behavior. The teachings of our parents or the policeman on the corner have great influence. We may all have participated in a Nurture versus Nature debate as to which factors are significant as to how humans behave. The nurture aspect of human behavior is well studied. Less appreciated by the general public is the role that our inherited genetic traits have on our behaviors. Fear, anger, aggression, greed, compassion, prejudice, morality and our tribal nature are among the traits that have a genetic basis.

Considering that we are an intelligent species, we still harbor some illusions. The one that many humans believe is that we are more than mere animals, we are a rational species, unlike other species. We also believe that we have free will, that we have complete freedom of choice, that there are no biological constraints upon our behavior. We just assume that we are rational and that, obviously, we have free will. Lovely thoughts; however, the scientific perspective is that we are tribal animals. This notion is not without controversy. It is a perspective that may be unpopular, unpleasant, perhaps even perceived as degrading. Indeed, we humans have progressed well beyond our primate ancestors, well beyond being hunter-gatherers. However, science tells us that we remain tribal animals. Therein lies the explanation as to our nature.

Humans did not arrive on planet earth as we now are. We share a history with all living organisms that live or have lived on this planet. We are one of over sixteen million species of plants and animals on the Tree of Life, a tree with millions of branches and twigs. Humans and our direct ancient ancestors have been on this planet for hundred of thousands of years. *Homo sapiens* are the survivors. We stand alone; we outwitted, outplayed and outlasted all other hominin species. For clarification, *hominid* refers to the group consisting of all modern and extinct Great Apes. This would include humans, chimpanzees, gorillas and orangutans plus all their immediate ancestors. Throughout this book I will use the more restrictive word *hominin*, which refers our group of modern humans and our ancient ancestors, back 6 million years. It does not include chimpanzees, gorillas and orangutans.

In ancient times, we handled unfamiliar phenomena with occult explanations. Today, explanations can be achieved with scientific inquiry. Human behavior is complex, almost mystifying, yet with the rapid advances in the sciences we have a greater understanding of humans and their behavior. It is our brain that controls our behavior. The brain is the centralized, purposeful processor and controller that exerts centralized control over the other organs of the body. While the brain is extremely complex, it has ancient roots. It has been well studied that inherited genetic traits determine how fearful we are, how aggressive or greedy we may be. There are genetic variants of these traits. Not everyone receives the same DNA. Not everyone is as fearful or greedy as the next person. It is our genetic traits that have ensured our evolutionary success. Our genetic traits were inherited from our parents and from our ancient ancestors, traits that have survival value. We seem to have noble behaviors as well as behaviors that could be considered ignoble. There are genetic links as to why some people are prejudiced, some are compassionate, even why some people hate Brussels sprouts. All these seemingly disparate behaviors have roots that go back millions of years to our ancient ancestors.

The last chapter will be an assessment of the nature of humans. For this assessment I will employ a concept first introduced to me by Dr. Garrett Hardin (1915-2003). When I was a student at University of California, Santa Barbara in the late 1950s, early 60s, he was a professor I greatly admired. Dr. Garrett Hardin was an ecologist, humanist and a popular lecturer. He was one of the early alarmists concerning overpopulation: *A finite world can support only a finite population; therefore, population growth must eventually equal zero.* He was also known for the *First Law of Human Ecology*, which states that we can never do merely one thing without consequences, everything is connected.

I remember that Dr. Hardin would occasionally challenge his students to think. "Be critically observant. Be rational and intellectually honest." To this end he introduced a different concept for observation. "What would a Martian think when observing some human behaviors?" The Martian would represent an entity that could be objective and free of human bias or prejudice. Of course, everyone knew that Martians were fictional, yet one student asked, "What does the Martian look like?" Garrett Hardin replied that one shouldn't consider the physical appearance of the Martian. Shape, color or gender are of no importance. Rather, he said, think of this Martian as a hypothetical super-intelligent life form, not from Mars, but a very distant place. This Martian can observe the behavior of humans with complete objectivity, and not be influenced by personal feelings, prejudice, bias, religious or political leanings. This Martian would have the facility to determine veracity. "Does the Martian have a name?" asked another fellow classmate. "This entity is one who will give a direct, complete and exact accounting. The entity represents the voice of reason, sound judgment and truth". Therefore, said Dr. Hardin, "We shall refer to this being as Parrhesiastes." The last chapter of this book will be an assessment on the nature of humans conducted by Dr. Hardin's Parrhesiastes.

*Parrhesiastes* is a derivation of *parrhesia*. Parrhesia can be defined as a verbal activity in which the speaker expresses his personal relationship to truth, and risks his life because he recognizes truth-telling as a duty to improve or help other people. Five essential characteristic traits of parrhesiastic speech are frankness, truth, danger, criticism and duty. Parrhesiastes is the one who uses parrhesia, speaks the truth, gives a complete and exact account of what he has in mind so that the audience is able to comprehend exactly what the speaker thinks. Its origins are traced to the Greco-Roman period in 6th and 5th century BCE. [1]

Fair notice: this assessment has a bias. Everyone has bias, myself included. I am a human. I am a member of the species of animals called *Homo sapiens.* I am programmed with my exclusive experiences and acquired knowledge and therefore I have a unique perspective. My inquisitive nature and Progressive proclivity have given me an idiosyncratic perspective of the human condition. This assessment is not necessarily better, not necessarily correct, but hopefully it will create some useful discussion.

# Two

## History

*History is a vast early warning system.*

—Norman Cousins (1915-1990), American political journalist, author, professor, and world peace advocate.

Previously I stated that our behaviors have genetic roots traceable back to our ancient ancestors. These early hominins didn't roam about on a placid planet. To appreciate the nature of humans we should review earth's geological history as defining influences on our ancestors.

When exploring the question as to when human history commenced, one can find various scenarios. Some religious people state that we came into existence only 10,000 years ago. Some anthropologists think that modern humans, that's us, evolved about some 70,000 years ago. Other anthropologists propose that we should consider that the human lineage commenced when our lineage split from chimpanzees about 6 millions ago. Biologists might consider that humans are the subsequence of an extremely long lineage that can be traced back to the origins of life itself, several billions of years ago. A more radical perspective is that one would consider that the beginning of humans commenced at the creation of the known universe. In 1973 Carl Sagan published The Cosmic Connection: An Extraterrestrial Perspective which included the following passage. "Our Sun is a second- or third-generation star. All of the rocky and metallic material we stand on, the iron in our blood, the calcium in our teeth, the carbon in our genes were produced billions of years ago in the interior of a red giant star. We are made of star-stuff."

With a nod to Carl Sagan, the history of humans begins with the creation of the known Universe, a Silent Big Bang, 13.8 billion years ago. Following was a period of 9.26 billion years before the earth was formed, about 4.54 billion years ago. For the first billion years of earth's existence, the planet was lifeless. At that time, the earth's atmosphere was composed primarily of methane, ammonia, water, hydrogen sulfide, carbon monoxide and carbon dioxide. Oxygen was either rare or absent.

Were these early earth conditions suitable for the creation of life? This was the question posed by two scientists. In 1953, Stanley L. Miller and Harold C. Urey, working at the University of Chicago, conducted a laboratory experiment which simulated the earth's early atmosphere. The Miller–Urey experiment consisted of water vapor, methane ammonia and hydrogen in a closed vessel. To simulate lightning, the contents of the vessel was continuously supplied with an electrical discharge. The result after one week was a slurry, consisting of some amino acids, the constituent parts of protein. [1]

There is abundant evidence of major volcanic eruptions 4 billion years ago, which would have released carbon dioxide, nitrogen, hydrogen sulfide and sulfur dioxide into the atmosphere. Subsequent experiments using these gases in addition to the ones in the original Miller experiment have produced more diverse biological molecules. [2]

An alternative explanation of the origin of life on earth is suggested by the unproven Panspermia hypothesis that life's origins may have occurred outside of Earth. Several billion years ago, water and inorganic chemicals for the creation of life likely existed on other planets such as Mars. In a Miller–Urey sort of conditions, primitive organic molecules, such as RNA, could have been synthesized on Mars. Mars and early Earth were bombarded by giant meteors.

Giant meteor impacts on Mars could have kicked up water and debris containing early life precursors and, by gradational forces, landed upon earth. [3,4] We often consider that life on earth is rather special. Life may be more universal than we suspect.

The earliest recognized life forms were prokaryotes, microscopic single-celled organisms without a distinct nucleus. These early life forms arose at least 3.6 billions years ago. The evolution of life did not take place peacefully on a placid planet. During the earth's past billions of years, the earth has gone through numerous thermal changes, extended periods of cold climate followed by warmer interglacial periods. Even the global location of the continents has drastically changed. The earth's lithosphere, the crust and upper mantle, is broken up into seven or eight major tectonic plates. The movement of the earth's tectonic plates has caused the formation and breakup of continents over time, including occasional formation of a supercontinent that contains most or all of the continents. The supercontinent Columbia formed about 1.9 to 2.0 billion years ago, and broke apart about 1.4 billion years ago. Another supercontinent, named Rodinia, formed about 1 billion years ago and lasted about 400 million years. About 600 million years ago, supercontinent Rodinia broke apart and re-assembled into another supercontinent called Pangaea. About 200 million years ago, Pangaea split apart into various land masses and drifted at very slow rates to what we now recognize as the present positions of the seven continents. The present rate of our tectonic drift is slow, very slow. Europe is drifting away from North America on tectonic plates at a rate of one inch per year. More than 25 million years ago, India, once a separate island on a quickly sliding piece of the Earth's crust, crashed into Asia. The two land masses are still colliding, pushed together at a speed of 1.5 to 2 inches a year. The forces have pushed up the highest mountains in the world, the Himalayas, and have set off devastating earthquakes. Recent earthquakes in Nepal are remnants of that collision. The concept of continental drift was first proposed in 1596 by Abraham Ortelius, a Flemish cartographer and geographer. He is also credited with the first modern atlas, the Theatrum Orbis Terrarum (Theatre of the World). The modern model of continental drift, based on plate-tectonics, was developed during the first few decades of the twentieth century and validated in the late 1950s.

The Earth has undergone at least five major Mass Extinctions, rapid and widespread decreases in life.

The *Ordovician-Silurian Extinction Event,* the earliest recognized extinction event, occurred 450-440 million years Before the Present (BP). 60% to 70% of all species perished.

The *Late Devonian Extinction Event* occurred 375-360 million years BP and 70% of all species disappeared. 300 million years BP, reptiles appeared.

The *Permian-Triassic Extinction Event*, 252 million years ago, was the largest event, extinguishing 96% of all species. Mammals made their appearance 220 million years BP.

The *Triassic-Jurassic Extinction Event*, 200 million years ago. 70% to 75% of all species went extinct. The age of dinosaurs was between 195 million years to 66 million years BP. Primates evolved about 75-85 million years BP.

The *Cretaceous-Paleogene Extinction Event* occurred 66 million years when a great asteroid crashed in the Gulf of Mexico, resulting in the demise of three-quarters of plant and animal species on Earth, including all non-avian dinosaurs. Some very small shrew-like primates species that dwelled subterraneously survived, feeding off the remains of decimated organisms. The small surviving very early primates diversified into such groups as lemur, tarsiers, monkeys and apes. The split of the chimpanzee lineage and the lineage to humans occurred about 6 million years BP. Anatomically modern humans appeared about 200,000 years BP.

Ever since the Precambrian period, 600 million years ago, ice ages or glacial ages have occurred at widely spaced intervals of geologic time—approximately 200 million years—lasting for millions, or even tens of millions of years. Glacial ages are dynamic periods with glacial periods alternating with warm periods, called interglacial periods. These cycles occur on a time scale of 40,000 to 100,000 years. The earth is presently in an interglacial period which started about 10-13 thousand years ago. When the earth enters a glacial phase, polar ice accumulates and atmospheric water condenses to form snow and ice at the polar regions of earth. Then the ice spreads out from the poles. Glaciation results in drier climate; tropical forests replaced by grasslands and eventually desertification. The most recent ice age, more precisely Glacial Age, began 2.6 million years ago. Hominin's evolution took place during this period, within a changing and challenging geological and climatic environment. *Hominin* is a primate of the family *Hominidae* that includes humans and their fossil ancestors. *Homo* is our genus and includes other members of the human clade after the split from the tribe *Panini* (chimpanzees). Our scientific name is *Homo sapiens.*

Several factors have been cited as causes for mass extinctions. Volcanic activity is associated with plate tectonics. Volcanic activity would have created hot magma, atmospheric particulates, sulfur oxides and carbon dioxide. The *Toba Super Eruption* in Sumatra that occurred about 69,000 to 77,000 years ago may have caused a global volcanic winter of 6-10 years. An ash layer, approximately 6 inches deep covered South Asia and the land bordering the Indian, Arabian and South China Seas. It has been speculated that the human population may have decreased to a few thousand.

Climate change is a significant factor that contributed to extinctions. Causes of various climate changes include a complicated dynamic interaction between such events as solar intensity, tilting of the earth's axis, eccentricity of the earth's orbit, the distance of the earth from the sun, position and height of the continents, ocean circulation, and the composition of the atmosphere as well as catastrophic events such as asteroid impacts and volcanic activity. The evolution of all life forms on earth took place over billions of years upon a volatile earth with shifting continents, drastic climatic changes to the atmosphere and inter- and intra-species competition.

13.794 billion years after the Big Bang, about six million years ago, the lineage leading to modern *Homo sapiens* separated from our ancestral near relative, the chimpanzees. Early man roamed a pristine planet. We now find ourselves on a planet that is less than pristine, suffering from overpopulation and a worsening climate. The rich are getting richer and the poor are growing in numbers.

## Evolution

*One of the most frightening things in the Western world, and in this country in particular, is the number of people who believe in things that are scientifically false. If someone tells me that the earth is less than 10,000 years old, in my opinion he should see a psychiatrist.*

—Francis Crick

Francis Crick, OM, FRS, Nobel Laureate (1916-2004) was a British molecular biologist, biophysicist and neuroscientist. Together with James Watson and Maurice Wilkins, he was awarded the Nobel Prize in 1962 for their discovery of the structure of DNA, deoxyribonucleic acid. From 1976 until his death, he was a Distinguished Research Professor at The Salk Institute for Biological Studies in La Jolla, California. His research focused on theoretical neurobiology and the study of consciousness. Francis Crick referred to himself as a humanist, which he defined as the belief that human problems can and must be faced in terms of human moral and intellectual resources without invoking supernatural authority. He publicly called for humanism to replace religion as a guiding force for humanity.

1925 was the year of a famous American legal case referred to as *The Scopes Monkey Trial.* John Scopes, a Tennessee high school teacher, was accused of teaching evolution in violation of the Butler Act, a Tennessee law prohibiting public school teachers from denying the Biblical account of man's origin. Arguing for the prosecution was William Jennings Bryan, a devout Christian and a leading American politician. Clarence Darrow, a leading member of the American Civil Liberties Union, argued for the defense. Scopes was found guilty and fined $100, but the verdict was overturned on a technicality.

Today, 91 years post this trial, many US states are still contentiously debating whether Evolution or Creationism and Intelligent Design should be taught in public schools. The public seems divided. There are several recent polls asking Americans their beliefs on evolution, with quite similar results. A Gallup poll found that 46% of Americans believe God created humankind in a single day about 10,000 years ago, a literal interpretation of the Bible. Just 15% say humans evolved without the assistance of God. The Gallup poll interviewed a random sample of 1,012 adults, aged 18 and older, living in all 50 US states and the District of Columbia. Another poll, by YouGov, a nonpartisan organization, found that 40% favor teaching creationism and intelligent design in schools. 21% said humans evolved without the involvement of God.

We think of ourselves as rational intelligent beings. This disparity between religious interpretations and confirmed science would substantiate the notion that we are not always rational and well-educated. This speaks to the inadequacies of our educational system and to the influence of religion in public policy. While school boards and religious groups grapple with issues of what sort of science, fiction or nonfiction, should be taught, I would like to refresh my readers on Evolution Theory, including the evolutionary history of *Homo sapiens*.

There are 8.7 million eukaryotic species on our planet. Eukaryotes have cells with nuclei and other membrane-bound structures. Not included are the prokaryotes, such as bacteria and viruses. It is estimated that tens of millions more species are yet to be discovered and described. Considering only reptiles, there are 10,038 recognized species. There are more than 22,000 species of orchids. 950,000 species of insects. Humans are just one of millions of other species on earth. [5]

The *Theory of Evolution* is a rational method to explain the great variety of species that have inhabited earth. The theory offers explanations as to how life forms arose, the lineages and the mechanisms that produce such diversity. The theory of evolution is the history of man and all organisms, past and present. The theory relies only upon evidence. It is devoid of moral or ethical principles. It does not offer direction nor a purpose of life. Religion is one, but not the only source, for moral codes and ethical guidance.

Charles Darwin, in his 1859 book On the Origin of Species, outlined the theory of evolution, a scientific theory that explains the diversity of life forms and the mechanism that created this diversity, *Natural Selection*. In his travels and his study of nature he observed that species produce offspring in numbers greater than just simple replacement, that of all the offspring each pair produced, only two on average survive. He also noticed that there was variation among the progeny, not all were absolutely identical. He was very familiar with the popular activity of plant and animal breeding, where enthusiasts would select for breeding a specimen with favored traits. He did not know about DNA, but he did understand that traits were inherited and that if an individual was born with a trait that offered a survival or reproductive advantage, even a very slight advantage, the statistical odds were that this trait would be passed on down through generations. He also observed that these modified traits appear randomly and that most changes were deleterious, that runts may appear in a litter. On occasion, a trait would arise that offered a slight biological advantage to that organism. From his geology knowledge, he knew that the environment is not static, that the earth's topography and climates were diverse and that both changed over time. With this broad knowledge, he postulated the encompassing theory of evolution and natural selection, nature's way of selecting the more fit individuals, *survival of the fittest.*

Darwin provided very strong evidence for man's common ancestry with the living apes, and that all human populations were more closely related to each other than to any living primate; they were the same species.

A hypothetical example of evolution: random mutations within a brood of chicks might give one chick a slightly longer beak, the ability to reach further down into a flower to extract just a bit more nectar. Another chick might have a slightly stronger beak, giving it the ability to crack open more seeds, or crack them open more quickly, and thus get more food per day. A small genetic change or mutation could assist that individual to survive, reproduce and thus pass on to future generations this mutation.

In one sense, all humans are alike. We all have hemoglobin, digestive enzymes to process our food, and opposable thumbs making grasping easy. But this is just too simplified. We are all different. While we all have the same proteins that make up our heart, our eyes and our brains, there can be variants in any protein or other biochemicals that are coded by our inherited DNA.

The structure of hemoglobin has been conserved over million of years, as it is superb for transporting oxygen to our red blood cells. The hemoglobin you have is virtually the same across all vertebrates. The structure of this molecule has been conserved, as it functions very well.

There is a caveat: DNA can mutate every time it is replicated. Mutations create novelties which may be selected as the environment changes. If a mutation is deleterious, the organism carrying this mutation may not survive and that mutated change is lost. If, however, that mutation has some, even a small, advantage for that organism, it may survive in that organism and be passed on to future generations. Any DNA sequence may undergo a mutation, substituting one base pair for the alternative. There are other possible DNA alterations that change the DNA sequence. Most mutations are deleterious, but a very few may offer that individual a competitive advantage; it may be more fit.

Commonly, and to a large degree, all the various biochemicals within a species are the same. Here is an example where there are variants within our species. Hemoglobin, the iron-containing protein within red blood cells, is responsible for oxygen transport to all the other cells within the body. It is basically the same molecule in humans and chimpanzees. There are, however, variant forms of hemoglobin. Sickle-cell anemia is a hereditary blood disorder characterized by an abnormality in the oxygen-carrying capacity of the hemoglobin molecule. This variant or mutant form of the molecule is identical to the normal molecule except at one specific location. Proteins are long sequences of amino acids. In the sickle-cell form of hemoglobin there is a simple change: the amino acid glutamine is replaced with another amino acid, valine. This very small but significant change in the amino acid sequence of hemoglobin causes the red blood cells to change shape, resembling a sickle. Additionally, the altered hemoglobin has a lower oxygen-carrying capacity and there are changes to the red blood cell's elastic properties. Normal red blood cells are quite elastic, which allows the cells to deform in order to pass through capillaries. The elasticity is lost in sickle-cell disease.

Individuals that are homozygous for this altered gene; they carry two copies of the altered hemoglobin suffer acute anemia and chronic health problems including an increased risk of death. Heterozygous individuals, those with but one abnormal copy have minor health problems because the normal allele, or gene, is able to produce over 50% of the hemoglobin. These carrier individuals may lack the vigor of non-carriers.

There is an adaptive advantage to individuals that are heterozygous; that is, they carried one sickle cell gene and one normal gene. Individuals that are heterozygous are immune to malaria, which is transmitted by mosquitoes. The malaria parasite, *Plasmodium,* spends part of its life cycle in mosquitoes and part in the red blood cells of humans or other animals. In a human that is a carrier of the sickle cell disease, the presence of the malaria parasite causes the red blood cells with defective hemoglobin to rupture prematurely, rendering the *Plasmodium* parasite unable to reproduce. Hence, in areas where malaria is a problem, people's chances of survival actually increase if they carry the sickle-cell trait.

Almost 300,000 children are carriers of the sickle-cell disease, mostly in areas with an abundance of mosquitoes: sub-Saharan Africa, the West Indies and South Asia. Many of the slaves that were brought to the Americas had this disease, and this fact may have given credence to the notion that these black people were, by nature, lazy. If they suffered from this anemia, they would have a lack of energy. By "acting" lazy, they may have outsmarted their slave owners; they wouldn't be expected to work as hard. Appearing lazy may have been self-serving.

One could say, with a stretch of imagination, that our great ancestors were yeast. What may be surprising to many is that humans have retained some genes that arose millions of years prior to humans emergence and have survived and still exist in humans. Baker's yeast, *Saccharomyces cerevisie,* is a single- celled organism used in the bread-making industry. Whereas the size of their genome is rather small compared to the human's genome, there are some surprising similarities between human and yeast genomes. Yeast has some 12 million base pairs arranged into about 6,000 genes. Humans, by comparison, have 3 billion base pairs arranged into 20 to 25 thousand protein-coding genes. And of those 6,000 yeast genes, about 2,000 are very similar to human genes. Because of the similarity of some human disease genes to those of some yeast genes, yeast is often the organism of choice in studying some genetic diseases. There have been over 50,000 published scientific articles describing yeast research. [6]

Considering the overwhelming scientific evidence that has been published in peer-reviewed journals, *Homo sapiens* did evolve from more primitive life forms and these more primitive forms themselves evolved from a primordial cocktail, the results of which were natural chemical reactions. The most learned scholars accept that we are the product of evolutionary processes and that we have retained some, maybe even a great abundance, of those traits that developed in earlier primates. If you honestly believe that The Theory of Evolution is *just a theory*, then hold out at arm's length an egg, palm side down. Open your hand; then tell me that Sir Isaac Newton's *Theory of Gravity* is *just a theory.*

## Hominins

*The Universe seems neither benign nor hostile, merely indifferent.*

—Carl Sagan (1934-1996)

Following are brief descriptions of a few of our close ancestral relatives. Presently there are about 20 known and described members of the tribe Hominin, the subfamily classification for members of our branch of primates as distinct from the tribe *Panini*, chimpanzee. These known ancestors most likely do not include the total of all hominin species that participated in the history of human origins. While the record is incomplete; research is ongoing. Paleoanthropologists study early human fossil teeth and skeletal remains, including cranial capacity, as well as found tools and evidence of tool-making, fossilized footprints, traces of campfires, bones of animals digested, and other food detritus. DNA evidence, computers and sophisticated mathematical models are used for tracking variables such as lineage, skull size or pelvic variations over time. Evolutionists estimate that humans separated from chimpanzees about 6 million years ago.

Paleoanthropologists may find the following simplistic and missing a great amount of rather significant detail. For ease of reading, I have simplified our ancestral history. What is important is that humans have a long and continuous history; there is no "Missing Link." The material presented is from multiple sources, both scientific literature and general paleoanthropology sites. [7]

Chimpanzees, *Pan troglodytes* and bonobos, *Pan paniscus,* are the closest living relatives to humans. Only the Zaire River separates the Chimpanzees from the Bonobos. The chimpanzee branch, not the bonobos leads towards humans, *Homo sapiens.*

Dr. Jane Goodall, a British primatologist, is best known for her 45-year study of social and family interactions of wild Chimpanzees in Gombe Stream National Park in Tanzania. She observed behaviors such as hugs, kisses, tickling, self-awareness and self-interest, pats on the back, and grief, traits that we would consider human.

Chimpanzees live in male hierarchical social groups, which can range from 40-60 individuals. A chimp community is organized more or less in linear fashion. It establishes social standing, with one male at the top, the "alpha" position dominating all females, although females have their own hierarchy, albeit much less straightforward. Minor disputes within the group are often handled with gestures and posturing rather than actual attacks. Occasionally, individuals switch from one group to another. Switching adds to a community's genetic diversity. Members of these groups develop strong affectionate bonds with each other that can last a lifetime. Mother-daughter bonds are very strong. Bonding has also been observed between siblings and pairs of males. Sex is strictly about reproduction, and reproductive tactics can include infanticide, the killing of offspring unrelated to a male chimp. Major disputes are often resolved by threatening displays or by fighting. Chimps are also capable of physical violence between tribes. Dr. Goodall observed a 4-year territory war between two groups of chimps that ended with one group killing all the other chimps in the other group.

In contrast to chimpanzees, the bonobos have some very unique characteristics. The bonobo society is marked by its easygoing ways, sexual equality, female bonding, and zeal for constant high-level sexual behavior amongst all members of a group. Bonobos apparently use sex to reinforce bonds within the group and to resolve conflict. Sex functions in conflict appeasement, affection, social status, excitement, and stress reduction. It occurs in virtually all partner combinations and in a variety of positions. This is a factor in the lower levels of aggression seen in the bonobo when compared to the common chimpanzee and other apes. Bonobos are perceived to be matriarchal and a male's rank in the social hierarchy is often determined by his mother's rank. [8] Worth considering is the question: If *Homo sapiens* had evolved from the bonobos rather than from chimpanzees, how might the world appear? Would the biosphere be pristine?

The similarity of DNA between humans and chimpanzees has been estimated to be between 95% and 98.5% depending on which nucleotides are counted and which are excluded. In other words, with respect to our shared DNA, we are very close to the apes. [9,10]

Progressing chronologically towards modern humans, the earliest known hominin is, *Ardipithecus kadabba.* This species lived about 5.6 million years ago in Eastern Africa and had a brain size of 300 and 350 cubic centimeters ($cm^3$), comparable to the chimpanzee. The human brain is about 1200 to 1400 $cm^3$. This species is known only from teeth, their wear pattern and smaller pieces of skeletal bones.

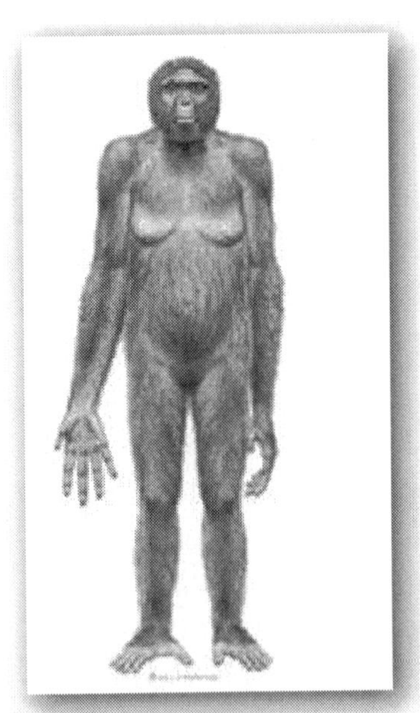

*Ardipithecus ramidus,* Scientific paleoartist Jay Matternes' rendition of Ardi. Wikipedia

Another species in this genus is *Ardipithecus ramidus,* who roamed Africa about 4.4 million years ago. Compared with chimpanzees, whose feet are specialized for climbing, *A. ramidus'* feet were better suited for walking. This species was about 120 cm tall (4 feet). The canine teeth of *A. ramidus* were smaller, and equal in size between males and females as compared with apes. From this, researchers infer certain aspects of the social behavior of this species: reduced male-to-male conflict, increased pair-bonding, and increased parental investment. Social behavior arose before enlarged brains and stone tool usage. In 1994, paleontologists unearthed in Ethiopia a relatively complete *A. ramidus* fossil skeleton of a female nicknamed *Ardi.*

*Kenyanthropus playtos,* known from a fossilized specimen, roamed Kenya between 3.5 and 3.2 million years ago. Some scientists suggest that this species may be an ancestor to the genus *Homo.* This species was geographically and temporally similar to another species, *Australopithecus afarensis,* but is physically distinctive. It is believed both occupied similar habitats, grassland and forested areas.

*Australopithecus afarensis* is a hominin that lived 3.85-2.95 million years ago. Found in Eastern Africa, this species is thought to be a direct or close relative of the lineage leading to *Homo sapiens.* The skull and brain are small, about the size of a chimpanzee, about 400 $cm^3$. They were mostly bipedal and found in the African savannas and arboreal environments. They likely expanded their diet to include meat, most likely from scavenging.

In 1974, in what is now Ethiopia, there was a remarkable find of several hundred pieces of bone representing about 40% of a female skeleton. This specimen was given the scientific name *Australopithecus afarensis* and classified as a member of the tribe *Hominin,* the taxonomic rank between family and genus. Estimated to have lived 3.2 million years ago, she was given the common name *Lucy,* after the Beatles' song *Lucy in the Sky with Diamonds,* which was played often by the anthropologists on that dig. She measured about 107 cm (3½ feet) tall and weighed 27-29 kg (60-65 pounds). The skeleton indicates evidence of Lucy being bipedal, walking on two legs. This species lived in trees as well as on land. This is evidence that bipedalism preceded increase in brain size. In 2013, Ethiopia expressed a desire to bring Lucy back. She now resides at the National Museum of Ethiopia in Addis Ababa.

A reconstruction of a female *Australopithecus afarensis* Wikipedia

Another scientific discovery in Kenya challenges conventional wisdom about human history: by whom and when, were stone tools invented? Stone tools were discovered near Lake Turkana that were crafted 3.3 million years ago. These tools would then be the oldest stone tools made by human ancestors. These are the "knives" of the Stone Age, stones that were intentionally flaked to have sharp edges. The Stony Brook team says the maker of these tools might have been an apelike creature called *Australopithecus,* or perhaps *Kenyanthropus.* That's hundreds of thousands of years before the genus *Homo* evolved. Lucy may have been the first toolmaker. [11-12]

*Australopithecus africanus* (not to be confused with *afarensis*) roamed southern Africa between 3.3-2.1 million years ago and is thought also to be in the lineage leading to modern humans. It is anatomically similar to *A. afarensis,* with a combination of human-like and ape-like features. It had long arms and asloping face, a slender build and a pelvis built for slightly better bipedalism than that of *A. afarensis.* It appeared to be a bit more similar to modern humans. The brain size ranged from 420 to 500 $cm^3$.

*Australopithecus sediba* fossil remains date to about 2 million years ago and it is unlikely that this species is in the direct lineage of modern humans. The brain has a volume of 420 $cm^3$ and their height was about 130 cm (4'3"). Its morphology suggests that this species may have been transitional between *A. africanus* and either *Homo habilis* or *Homo erectus.*

At the time of the last glacial age, 2.6 million years ago, well after the divergence of humanoids from chimpanzees, a gene that is associated with brain cell growth was mutated to an altered form. This altered gene, SRGAP2, helped our brain cells grow faster and make more connections, enabling the brain to become more complex. This mutation was a significant event in the evolution of hominins. [13]

Some anthropologists argue that there may have been as many as ten species within the genus *Homo.* Presented here are a few of the more recognized species.

Fossil studies have provided evidence that *Homo* evolved from *Australopithecus.* In 2013, in Afar, Ethiopia, a jawbone was discovered and attributed to *Homo habilis*, the first of the genus *Homo.* It was dated to 2.8 million years ago. This fossil combines the morphology observed in later *Homo* with primitive traits seen in early *Australopithecus.* The distinguishing feature of these two species, *Australopithecus* and *Homo*, is that *Homo* species had smaller jaws and teeth, indicating that stone tools were used to grind and process food. This was at a time when the environment had changed from forest to grassland. [14]

*Homo naledi* is also one of the earliest species of the genus *Homo.* Specimens have recently been discovered in the Dinaledi Chamber within the Rising Star cave in the northwest of Johannesburg, South Africa. This species had a mixture of anatomical characteristics of primitive species as well as characteristics of modern humans. The hands of *H. naledi* appear to have been better suited for object manipulation than those of Australopithecines. Males stood about 150 cm (5') tall, females smaller. The jaw, teeth and feet of this species are similar to modern humans, whereas the brain was smaller, about 450 - 550 $cm^3$. As this species is newly discovered there is still some uncertainties as to its place in the phylogenetic history of the genus *Homo* as well as its age, now estimate to have lived about 2.5-2.8 million years ago. [15]

*Homo habilis* or "Handy Man" was so named because of evidence of stone tools found with its remains. This species was short and retained some ape-like morphology, its arms nearly as long as its legs. *H. habilis* seemed to lack hunting skill and relied upon scavenging recently killed large mammals, as animal bones uncovered showed carnivore marks as well as man-made marks, indicating the use of primitive stone tools to dismember the previously killed animals. Tools could have been used to process vegetation for easier digestion as well at to crush the bones for the marrow. *H. habilis* may have been the prey of sabre-tooth cats that roamed Africa 5 million to1.2 million years ago.

Reconstruction of Homo habilis
Westfälisches Museum für Archäologie, Herne, Wikipedia

*Homo ergaster. Working Man* is also a direct ancestor. He lived in eastern and southern Africa between 1.8 and 1.3 million years ago. He was long-legged, able to walk longer distances. Females grew to about 157 cm (62") and males reached 178 cm (70"), with a brain size of about 860 $cm^3$. Their bodies may have been relatively hairless as a way of improving body cooling by sweating. *H. ergaster* may have been the first to have a human-like voice, to harness fire, whether by obtaining it from natural occurrences or by igniting artificial fire. Their tools were mainly used on meat, bone, animal hides and wood. There is no archaeological evidence that *H. ergaster* had symbolic thought such as figurative art or burying their dead. Fossils found in Eurasia in the Republic of Georgia may represent the earliest evidence for the emergence of early humans from Africa into Eurasia 1.75 million years ago. *Homo ergaster* is considered to be the common ancestor of two groups of humans that took different evolutionary paths. One of these groups was *Homo erectus;* the other group ultimately became our own species *Homo sapiens.* [16]

A common known ancestor is Homo erectus, Upright Man. The earliest fossils date to around 1.9 million years ago, the most recent to around 143,000 years ago. This species originated in Africa and some members, about 2 million years ago, migrated out of Africa and spread as far as Georgia, India, Sri Lanka, China and Java. Fossils discovered vary and therefore there are still ongoing debates as to the scientific classification of this species. Adults were slender and weighed 40-68 kg (90 to 150 lbs). They stood slightly over 150 cm (5') to nearly 180 cm (6'), evidence of a large variation of skeletal size variation. Brain size varied over its habitat range and chronology; but the latest Javan specimens measure up to 900 cm$^3$. The stone tools they manufactured were used as a scraping or cutting tool or possibly they could have been used in hand-to-hand combat or tossed at foe or prey. It has been suggested that H. erectus may have been the first hominin to use rafts to travel over oceans. Their disappearance in Asia may have been caused by the Toba Super-eruption in Indonesia 69,000 to 77,000 years ago. This volcanic eruption caused a worldwide volcanic winter lasting 6-10 years. [17]

A model of the face of an adult female *Homo erectus.* Reconstruction by John Gurche, Smithsonian Museum of Natural History, based on KNM ER 3733 and 992. Wikipedia.

*Homo antecessor*, *Pioneer Man,* discovered in Spain, dates to 1.2 million to 800,000 years ago. Rather tall, 168 cm (5'6") to 183 cm (6') and weighing about 90 kg (200 lbs), however their brain was slightly smaller than modern man, about 1,100 cm$^3$. Fossil remains and tools discovered in England dating to 950,000 years ago indicates that *H. antecessor* was the earliest *Homo* in Northern Europe.

*Homo heidelbergensis* was first discovered near Heidelberg, Germany in 1907. It appears intermediate between *Homo erectus* and fully modern humans. *H. heidelbergensis* may not have been a direct ancestor of *H. sapiens.*

*Homo neanderthalensis* or Neanderthal was discovered in 1856 in the Neander Valley in Germany. They are often thought of as our brutish, dimwitted ancestors that lived in caves. They did occupy caves at times, they were stronger than us, but they may not have been all that dimwitted. However, they are not our direct ancestors. [18]

A reconstruction of *Homo heidelbergensis*. Wikipedia

Males stood about 168 cm (5'6"), weighing about 77-83 kg (170-185 lbs), and females were about 152 cm (5') tall and weighed about 66kg (145 lbs). Some genetic studies have suggested that Neanderthals may have had red hair and light skin color. [19] An analysis indicates that Neanderthals' brains were similar or larger in size to modern man's, up to 1600 cm$^3$. Fossil data suggests, however, that their brain structure was rather different. The results show that larger areas of the Neanderthal brain, compared to the modern human brain, were given over to vision and movement, probably as a consequence of living in northern climes. This left less room for the higher level thinking required to form large social groups and for extensive creativity. Neanderthals were an advanced species, capable of intelligent thought processes and were able to adapt to and survive in some of the harshest environments known to humans. They were advanced tool makers. They used fire, made clothing and lived in large complex social groups and occupied caves. Their diet consisted of both large animals and vegetation. [20]

Neanderthals and *Homo sapiens* likely shared some elements of speech and language a half million years ago. Two scholars, Dan Dedie and Stephen C. Levinson from the Max Planck Institute for Psycholinguistics, evaluated a broad range of evidence from linguistics, genetics, paleontology, and archaeology. They argue that recognizably modern language is most likely an ancient feature of our genus, predating at least the common ancestor of both modern humans and Neanderthals. [21]

The lineages leading to modern humans and Neanderthals likely diverged in Africa about 600,000 years ago from a common ancestor, most likely, *Homo heidelbergensis.*

This common ancestor later split into two parallel branches. One branch became modern man, *Homo sapiens,* about 200,000 years ago and remained in Africa until 47,000–65,000 years ago. The other branch migrated out of Africa to Europe and beyond about 230,000 years ago. This branch later split again into two lineages, Neanderthals and Denisovans.

Neanderthals lived in Europe and western Asia with a range that extended as far east as Siberia and as far south as the Middle East. The overlap of Neanderthals and modern humans in space and time suggests the possibility of interbreeding. However, there is ongoing rigorous debate whether Neanderthals and *Homo sapiens* did interbreed. When scientists discovered that modern humans shared some genetic sequences of DNA with Neanderthals, their best explanation was that, at some point, the two species must have interbred. Scientists of the Neanderthal Genome Sequencing consortium suggest that interbreeding may have occurred when modern humans carrying Upper Paleolithic (between 50,000 and 10,000 years ago) technologies encountered Neanderthals as they expanded out of Africa, most likely 47,000–65,000 years ago. [22]

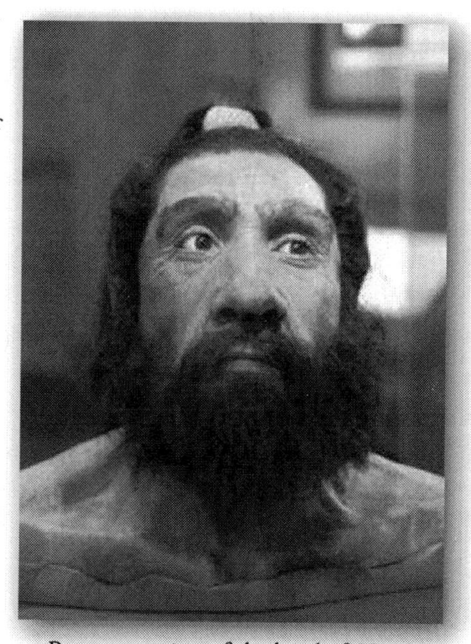

Reconstruction of the head of *Homo neanderthalensis,* Shanidar 1 fossil, a Neanderthal male who lived c. 70,000 years ago (John Gurche 2010). Wikipedia

A recent scientific study has indicated that modern humans interbred with not only Neanderthals but also with another closely related hominin species, the Denisovans at least four times tens of thousands of years ago. It is conjectured that this interbreeding may have given modern humans immunity to additional pathogens. It is certain that the lineage of *Homo sapiens* is complicated and the complete story is unfolding with additional research. [23]

The Denisovans, classified as a sub-species of *Homo sapiens,* are known from the DNA of a 80,000-year-old bone fragment discovered in a cave called Denisova cave, located in the Altai Mountains of Siberia, Russia. Researchers were able to extract both mitochondrial and nuclear DNA and discovered that the Denisovans are a sister group to the Neanderthals. The Denisovans also managed to cross the Wallace Line, an oceanic barrier formed by a powerful biogeographic marine current that marks the division between European and Asian mammals to the west from marsupial-dominated Australasia to the east. DNA evidence shows that Denisovans bred with ancestors of the present-day New Guineans and other Melanesians. [24]

Discoveries at the site of La Chapelle-aux-Saints cave, bordering the Sourdoire Valley, France, suggests the possibility that Neanderthals intentionally buried their dead, at least 50,000 years ago, before the arrival of anatomically modern humans in Europe. These discoveries support the contention that these Western European Neanderthals possessed complex symbolic behavior. [25]

One of the last outposts of Neanderthals was the extensive cave system called El Sidrón, discovered in Northern Spain. In the El Sidrón cave, researchers found the fossilized remains of a group of Neanderthals who lived and perhaps died violently, approximately 41,030–39,260 years ago. The bones had cut marks that were made by the blow of a stone tool. The inference is that these Neanderthals were cannibalized for their bone marrow and brains. The reason could have been ritual or starvation or they could have become victims of violent acts. In another Neanderthal cave, farther south on the Rock of Gibraltar, researchers discovered stone spearpoints and scrapers. This cave was occupied from 125,000 years ago until their disappearance 40,000 years ago. [26]

Neanderthals occupied a large range and their disappearance occurred in pockets at different locations and different times, which coincided with climatic changes as well as the appearance of *Homo sapiens.* There was an overlap with *Homo sapiens* that lasted between 2,600 to 5,400 years. [27]

The many factors suggested for the Neanderthal's demise include their lack of fitness to the warming environment and the organization of their brain, which may have lessened their social, competitive and creative skills. What is acknowledged is that there was overlap with *Homo sapiens* and Neanderthals, both temporally and geographically. There was a transitional period when there was ample time for the possible transmission of culture, symbolic behaviors and or genes. Did *Homo sapiens* witness this extinction of Neanderthals? Or did they participate? Both seemed to have occupied the same niche. Both these hominins most likely hunted the same animals for food. A niche includes an organism's habitat, its food sources and, simply stated, its occupation. Garret Hardin's *The Competitive Exclusion Principle* states that no two non-interbreeding species can simultaneously occupy the same ecological niche and geographical territory. Over time, one species will out-compete the other. This could include out-breeding or elimination. [28]

*Cro-Magnon* is a common name that has been used to describe the first early man. However, it has no formal taxonomic status, as it refers neither to a species nor subspecies nor to an archaeological phase or culture. The name comes from the combination of the local language for rock shelter or cavity, Cro, along with the name of the owner of the land Magnon, where the first specimen was found in southwestern France. *Cro-Magnon* were the artists linked to the well-known Lascaux cave paintings.

## *Homo sapiens*

*Man is physically as well as metaphysically a thing of shreds and patches, borrowed unequally from good and bad ancestors, and a misfit from the start.*

—Ralph Waldo Emerson (1803-1882)

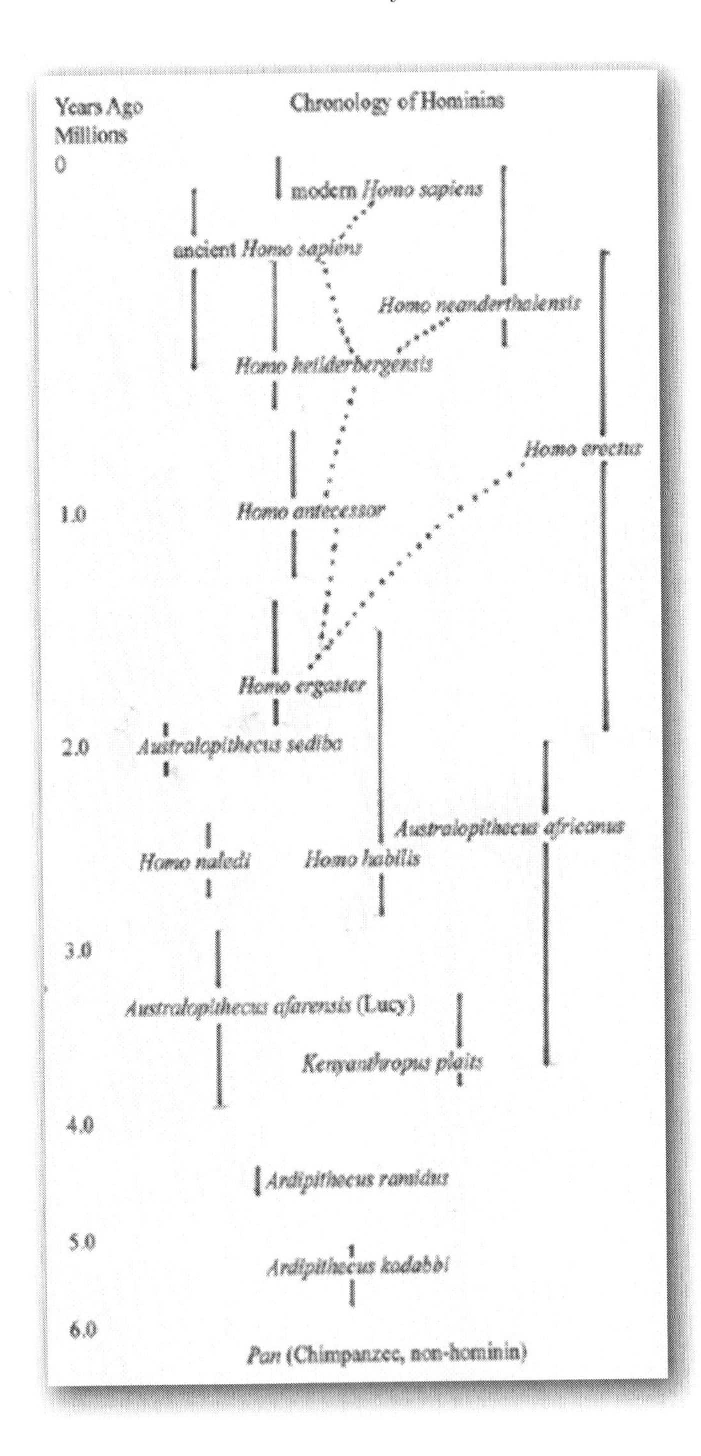
Years Ago
Millions
Chronology of Hominins
0
modern Homo sapiens
ancient Homo sapiens
Homo neanderthalensis
Homo heliderbergensis
Homo erectus
1.0
Homo antecessor
Homo ergaster
2.0
Australopithecus sediba
Australopithecus africanus
Homo naledi
Homo habilis
3.0
Australopithecus afarensis (Lucy)
Kenyanthropus plaits
4.0
Ardipithecus ramidus
5.0
Ardipithecus kadabbi
6.0
Pan (Chimpanzee, non-hominin)

I presented, in chronological order, the names and brief characterizations of just a few of our ancestral parents and cousins as a glimpse of the evolutionary path towards modern man. The lineage of hominins up to modern humans is not a straight line. There were numerous small- brain, upright hominin species in the first few million years after the divergence from the chimpanzees. It remains difficult to know with exactness the path leading to *Homo sapiens*. Genes may have flowed between species. Some species may have been a dead-end lineage due to being unfit for changing interglacial warming and cooling periods, the environment, or some species may have been extinguished by predators or by other hominin species. Interbreeding between early species may have occurred. Species of the genus *Homo* became diverse morphologically in different localities. Scientists have discovered specimens in the same location and dating to the same time, but having variable morphological traits.

The rarity of skeletons makes the reconstruction of body size and shape dependent on many assumptions, which can be subject to interpretation. Another limitation arises from homoplasy, the appearance of similarities in separate evolutionary lineages. Homoplasy was common in hominin evolution. Some physical traits are independent evolutionary novelties. Other physical traits are not independent. Flexion of the base of the skull and being flat-faced are generally correlated, as is jaw size and tooth size.

As a consequence of these confusing finds, paleoanthropologists can be grouped into either Splitters or Lumpers. These terms are terms used by taxonomists to characterize two groups of biologists. Lumpers group specimens having somewhat similar characteristics into one large diverse species. The Splitters believe that a group of organisms with some diversity may represent more than one species. Paleoanthropologists are no exceptionThere is controversy revolving around an acceptable definition of what exactly is a species. The simple and traditional definition of a species is a group of organisms consisting of similar individuals capable of interbreeding and producing viable and fertile offspring. The horse and donkey's offspring, the viable but sterile mule, is evidence that the parents are of two species. Dogs, with a large diversity of breeds, are of one species. Offsprings of the different breeds meet the criteria of one species, *Canis domesticus.* Unfortunately, the archeological tools available cannot determine the viability and fertility of any offspring that may have occurred between diverse hominins. What can be stated with certainty is that a more accurate accounting of the lineage leading to man will change as new discoveries and research continue. One can also confidently say that any change in this history of man will be in the minor details, that the present survey is reasonably accurate.

Incidentally, the scientific binomial nomenclature, using the taxonomic system of genus and species, was developed by Carolus Linnaeus (1707-1778), a Swedish botanist, zoologist and physician, as a system for describing and ranking all organisms. The binomial nomenclature is still used today. Linnaeus is known as the father of taxonomy.

The evolution of life occurred over billions of years on a planet that was constantly changing. There were shifting continents, volcanic activity, asteroidal impacts and drastic changes in temperatures. As the environment changed, organism either had genetic modifications suitable for the new environment or they became extinct. From six million years ago to about one million years ago there was a gradual cooling of the planet. Polar regions accumulated ice and equatorial regions became drier, with forests becoming grasslands. From one millions years to present times there were numerous small fluctuations in surface temperature resulting in shifts in flora and fauna distribution. Within this challenging environment, the genus *Homo* arose evolving from chimpanzees-like animals to morphological and intellectual modern humans.

*Homo sapiens* originated in Africa and obtained anatomical modernity 200,000 years ago and began to exhibit full behavioral modernity around 50,000 years ago. While other hominins migrated out of Africa 2 million years ago, *Homo sapiens* migrated out of Africa about 70,000 years ago. We were hunter-gatherers until about 10,000 years ago. Cooperative behavior, as distinct from the fierce aggression observed between chimp groups, was their norm.

In a nod to the CBS television series Survivor *Homo sapiens* outwitted, outplayed and outlasted all other species in our genus.

For Those who would enjoy a wider view of human evolution, Lone Survivors: How We Came to be the Only Humans on Earth, by paleoanthropologist Chris Stringer would be an excellent choice. as well as *Sapiens:* a Brief History of Humankind by Yuval Noah Harari.

# Three

## Brain

*The brain is more than an assemblage of autonomous modules, each crucial for a specific mental function. Every one of these functionally specialized areas must interact with dozens or hundreds of others, their total integration creating something like a vastly complicated orchestra with thousands of instruments, an orchestra that conducts itself, with an ever-changing score and repertoire.*

—Oliver Sacks (1933-2015), Professor of Neurology at New York University School of Medicine. Many of his books have been best sellers, including Awakening (1973), The Man Who Mistook his Wife for a Hat (1985) and Musicophilia: Tales of Music and the Brain (2007).

Brain research is one of the most exciting area of science today. The United States and the European Union have launched new programs to better understand the brain. The National Institutes of Health spends $4.5 billion a year on brain research, much of it directed towards research on diseases like Parkinson's and Alzheimer's. In April 2013 President Obama announced a broad new research initiative starting with $100 million in 2014 on the *Brain Initiative,* to invent and refine new technologies to understand the human brain. [1]

Harvard's *Human Connectome Project,* funded by the NIH, is a $40 million five-year effort to produce a structural map of the mouse brain at the level of magnification that shows packets of neurochemicals at the tips of brain cells. *The Salk Institute* in La Jolla plans to spend a total of $28 million on new neuroscience research. *The Allen Institute for Brain Science* in Seattle employs the disciplines of math, physics, engineering, systems-level and molecular neuroscience, molecular biology, genetics, genomics and information technology. Dr. R. Clay Reid, a senior investigator at the Allen Institute, said two fundamental problems must be solved. The first is "How does the machine work, starting with its building blocks, cell types, going through their physiology and anatomy." Second, "How does that neural computation create behavior? How does the mouse brain decide on action based on that input?" Universities and private research institutions in the US and abroad are also contributing to brain research. [2]

The brain is the centralized purposeful processor and controller that exerts centralized control over the other organs of the body. Our brain is that organ that is responsible for our high levels of abstract reasoning, language, problem-solving, culture, behavior, moods and emotions, all arising from electrical impulses in the brain, all working in perfect harmony. The human brain could be compared to a symphony. When all is going well, beautiful music. But as you can imagine, things can go wrong. Hopefully all the musician's instruments are well tuned, the musicians are following the same score and paying attention to the conductor. There are so many factors; neurons, biochemicals and associated structures in our brain that can and do, sometimes, go astray.

That which is critical to appreciate is that the brain is ancient. The chronology extends back at least 500 million years. From the first interconnected neurons to our highly developed processor, each small change in the developmental history of the brain was selected if it had survival value. The brain we have today is not too distinct from the brain of earlier hominins. The genetic behavioral traits of our ancient *Homo* species were, for the most part, conserved and are displayed by modern humans.

> The Astonishing Hypothesis is that, *You, your joys and your sorrows, your memories and your ambitions, your sense of identity and free will, are in fact no more than the behavior of a vast assembly of nerve cells and their associated molecules. As Lewis Carroll's Alice might have phrased it: You're nothing but a pack of neurons.*
>
> —Francis Crick [3]

The assemble of nerve cell and their associated molecules that Francis Crick referred to is commonly called our brain. A little simplified refresher course may be in order. The adult human brain weighs about 1.4 kg (3 lbs) with a volume of around 1350 $cm^3$, about 3.6 times larger than the ape's brain. In the human brain the cerebral cortex contains an estimated 86 billion neurons. Widely separated brain regions communicate through the long-range projection of neurons called axons. The number of intra-neuronal connections exceeds 100 trillion. By comparison a cockroach has about one million neurons while the Chimpanzee has about 6.7 billion neurons. [4]

The *Cerebral Cortex*, the thin 1.5 to 5 mm outer layer of the brain, is divided into two symmetrical halves. It comprises what we know as the grey matter. The cerebral cortex play a key role in memory, attention, perceptual awareness, thought, language and consciousness. It is connected to and communicates with various subcortical structures such as the thalamus and the basal ganglia.

The *Cerebrum* is the largest part of the brain, occupying 2/3 of the brains mass. It is divided into two hemispheres connected by the corpus callosum.

The *Limbic System* is a complex set of evolutionarily primitive brain structures located on both sides of the thalamus, right under the cerebrum on top of the brainstem. It is not a separate system but a collection of structures.

Limbic system structures are involved in many of our emotions and motivations, particularly those that are related to survival. Such emotions include fear, anger, and emotions related to sexual behavior. Two large limbic system structures, the amygdala and hippocampus play important roles in memory. The limbic system is responsible for a variety of functions including adrenaline flow, emotion, behavior, motivation, long-term memory and olfaction. The term Limbic System is a historical concept and may not be as accurate as neuroscientists may presently use however, it will suffice here.

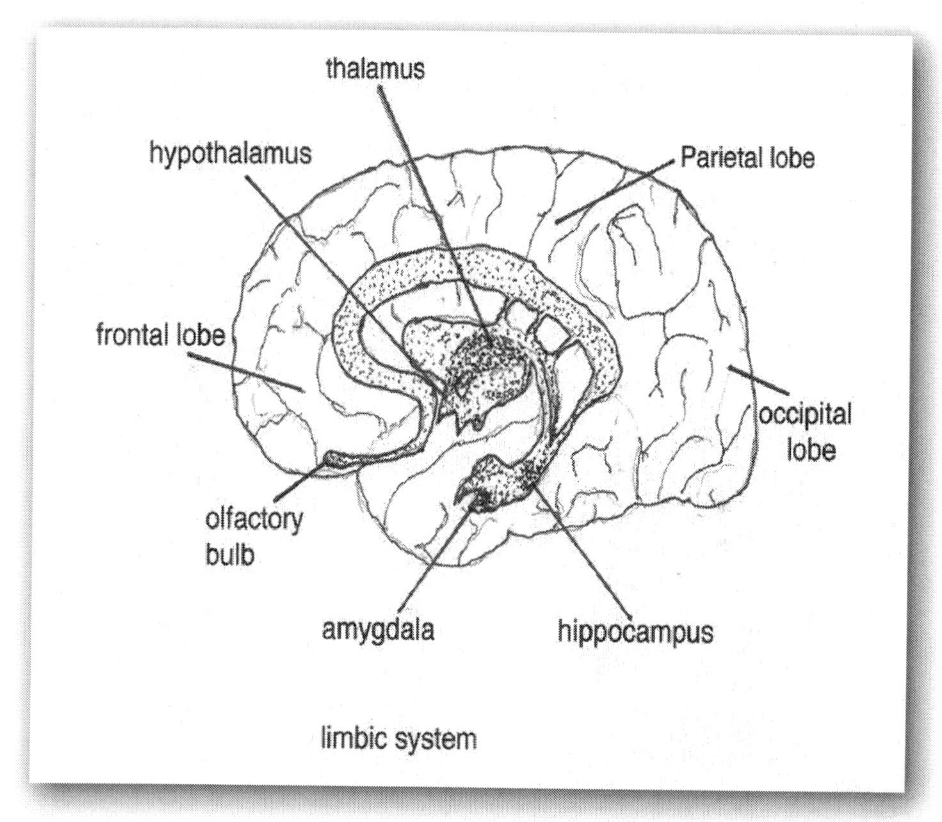


limbic system

The *Amygdalae* (plural) are two small almond-shaped groupings of cells deep within both sides of the brain. Sensory information from the outside, such as visual, olfactory, auditory, touch and pain converge in the amygdala. The amygdala is responsible for determining what memories are stored and where the memories are stored in the brain dependent upon the emotional strength an event invokes. As part of the threat-detection system, the amygdala connects sensory inputs with all the systems involved with emotional reactivity. The amygdala becomes activated when an individual expresses feelings of fear or aggression, particularly those that are related to survival. In addition to external sensory signals, the amygdala connects with higher order systems such as the prefrontal cortex. The amygdala integrates all information and makes associations, such as food and bell-ringing, as in Pavlov's experiments with dogs, and then stores the information as associated memories as part of the processing of memory, decision-making and emotional reactions. The amygdala doesn't just respond to what we see out there in the world but rather to what we imagine or believe about the world. [5]

The *Thalamus* is a symmetrical structure of two halves situated between the cerebral cortex and the midbrain. It functions as a processor of information receiving input from eyes, ear, spinal cord and relays sensory and motor signals to the cerebral cortex. The thalamus functions as the regulator of consciousness, sleep, and alertness.

The *Hypothalamus,* about the size of an almond, is located just below the thalamus. The hypothalamus produces sex, growth and stress-related hormones and links the nervous system to the endocrine system via the pituitary gland. It is involved with basic functions: sex, temperature control, sleep, attachment behaviors, circadian rhythms and has a regulatory role in aggression.

The *Hippocampi* (plural) are two structures on either side of the brain. The hippocampus plays important roles in the consolidation of information, from short-term memory to long-term narrative memory and spacial navigation. In Alzheimer's disease, the hippocampus is one of the first regions of the brain to suffer damage. Memory loss and disorientation are included among the early symptoms. The hippocampus sends memories out to the appropriate part of the cerebral hemisphere for long-term storage and retrieves them when necessary. Damage to this area of the brain may result in an inability to form new memories…

The *Olfactory Bulb* is a multi-layered cellular structure that transmits smell information from the nose to the brain.

The *Temporal Lobes* are involved in the retention of visual memories, processing sensory input, comprehending language, storing new memories, emotion, and deriving meaning.

The *Occipital Lobe* is the visual processing center of the brain and includes most of the anatomical region of the visual cortex. It is the central processor for visual reception, visual-spacial processing, movement and color recognition.

The *Frontal Lobe* is associated with attention, abstract thinking, behavior, problem-solving tasks, physical reactions and personality. It contains most of the dopamine-sensitive neurons in the cerebral cortex. The dopamine system is associated with reward, attention, short-term memory tasks, planning and motivation. Dopamine tends to limit and select sensory information arriving from the thalamus to the forebrain. During human evolution the frontal lobe's expansion is linked to the our intellectual advancement. The increase in neural connections would be necessary for generating abstract symbolic concepts and planning tasks. The evolution of human complex functional neural organization may have been a long-term process, involving at least a million years. [6]

There is also the view that modern neural organization is the result of a relatively sudden genetic mutation that took place in populations from Africa only 50,000 years ago. [7]

The *Parietal Lobe* processes and integrates sensory information. Major sensory inputs, such as taste, touch, sound, pain and temperature are relayed through the thalamus to the parietal lobe.

## Neurotransmitters

It is well known that susceptibility to certain diseases or afflictions has genetic links. There is a diversity of genetic differences that can lead to various physiological and psychological conditions. Some individuals with a specific genetic allele may be more susceptible to schizophrenia than others without that allele. There are genetic tests available to physicians to aid in the diagnosis and therapy for more than 1000 different diseases. There are a variety of home tests kits available, ranging from testing for breast cancer alleles to mutations linked to cystic fibrosis.

Less appreciated is that many of our behaviors also have a genetic link. Of course, major outside influences affect behavior such as one's experiences, teachings, cultural values, ethics, authority and coercion. Scientists are finding genetic links to many behaviors and publishing their results in the scientific literature. It will be the nature rather than the nurture aspect that will be our focus of human behaviors.

Whenever an event is consciously perceived by an individual, a response of some sort is likely. This reaction could be fear, anger or aggression or it could invoke a response of love or compassion, dependent upon one's assessment. Communication of information along the complex and highly regulate neuronal system is accomplished by the movement of chemical messengers called *neurotransmitters*. Neurotransmitters are the brain's chemicals that communicate information throughout our brain and body. Neurotransmitters transmit signals from one neuron, across a synapse, to the adjacent neuron where it will bind as a specific site. This action will then elicit either an inhibitory or excitatory signal. [8,9]

There are numerous factors in the complex signaling pathway which determine how well a neurotransmitter functions. The actual DNA sequence of a gene coding for a neurotransmitter or any of the other components of the signaling pathway can vary one individual to another. There are numerous processing steps from a specific segment of DNA that is the code for a protein all the way to the actual functionality of that gene.

Genes have *promoters*, sequences of DNA that switch on and off the gene's transcription. Not all genes are switched on all the time, only when appropriate.

*Transcriptional* steps occur when DNA is transcribed into single stranded RNA called messenger RNA (mRNA). This molecule, mRNA, transports the genetic message from the DNA to the protein-making machinery of the cell.

Next in the processing procedure are *translational* steps which chemically translated the RNA into a functional protein.

Neurotransmitters, once synthesized, are transported along the signaling path way by specific *transporter* proteins.

For functionality, specific *receptor* proteins bind neurotransmitters.

And on top of all this there are other considerations. Functionality of the neuronal signaling system is dependent upon numerous factors: the *affinity* of relevant biochemical to their target, the *rate* of transport and the *level* of expression of the various mentioned proteins. (There are other factors, intentionally omitted.) Altering aspects of neurotransmitter release, binding, and re-uptake or removal by pharmacological or other means is central to many therapeutic strategies. [9]

Variants (or mutations) in any of the numerous neurotransmitters or in factors of the signaling pathway can alter functionality. A few of us express behaviors in excess and others have behaviors that seem to be subdued. Nurturing could be a factor but the role of genetics offers additional explanations.

This is the important take-home lesson: human behavioral traits have a genetic component; that is, the trait and all the components for expression are encoded within one's DNA. Some traits evolved millions of years ago as these traits had survival values. Within the complexity of the neuronal signaling pathways, genetic mutations may occur within individuals and these mutations can be passed on through generations, if they have survival value and, of course they are not deleterious mutations. We know that some behavioral traits run in families. And finally, the expressed level of a trait may differ between individuals; not everyone has the same genetic code.

Scientists do not yet know exactly how many neurotransmitters exist but more than 100 biochemical messengers have been identified. These biochemical agents can be classified into two broad categories: small-molecule neurotransmitters and larger neuropeptides. Small-molecule neurotransmitters mediate rapid synaptic actions whereas neuropeptides tend to modulate slower ongoing synaptic functions.

Attention will be given to a few common neurotransmitters.

*Serotonin* is small mood-regulating neurotransmitter molecule, molecular weight (MW) = 176. It functions as a regulator of mood, appetite, and sleep. It also has cognitive functions including memory and learning. Serotonin is associated with the feeling of well-being and happiness. In humans, sexual and impulsive behavior, depression, fear, anxiety and aggression are, at least partially, an effect of the decrease in serotonin.

Low serotonin levels weaken communication between the amygdala and the frontal lobes, and this results in lower control of emotional responses to anger. When serotonin levels are low, it may be more difficult for the prefrontal cortex to control emotional responses to anger that are generated within the amygdala.

Serotonin is widespread throughout the Animal Kingdom from amoebas to the venom of insects to humans. In Macaque monkeys it has been found that alpha males have twice the level of serotonin in their brain than that found in subordinate males and females. Serotonin is also distributed throughout the Plant Kingdom. It is found in the seeds and fruits of many flowering plants where it may serve to stimulate the digestive tract of ingesting animals into expelling the seed, one of the ingenious schemes used by plants to distribute seeds.

*Dopamine,* an intercellular messenger (MW = 153), is in the nervous system of virtually all multicellular animals as far back as 500 million years ago. In humans it is a hormone and a neurotransmitter. In the brain, dopamine functions as a neurotransmitter. Dopamine plays a role in behavior, cognition and in reward-motivated behavior. Rewards such as food and sex as well as addictive drugs amplify the effects of dopamine. High levels of dopamine are associated with heightened levels of impulsive behavior, anger and aggression.

Drugs that increase the effects of dopamine such as cocaine or amphetamine, produce heightened levels of impulsive behavior. Low levels of dopamine lead to torpor and slowed reactions. Schizophrenia involves altered levels of dopamine activity and antipsychotic drugs have their primary effect by attenuating dopamine activity.

Many plants, including a variety of food plants synthesize dopamine to varying degrees. The highest concentrations are found in bananas and in lower detectable concentrations in potatoes, avocados, broccoli and Brussels sprouts. There is evidence that dopamine plays a role in the response to stressors such as bacterial infection. Dopamine can also act as a growth-promoting factor in some situations by modifying the way that sugars are metabolized.

*Oxytocin* (MW = 1007) is a mammalian hormone that also acts as a neurotransmitter in the brain. It is made in the hypothalamus and transported to and secreted by the pituitary gland. Sometimes called the *Bonding Hormone,* oxytocin acts primarily as a neuromodulator in the brain. It plays a key role in the regulation of social cognition and behavior. This hormone serves an important role in sexual arousal, orgasm, social recognition, ethnocentric behavior, male-female bonding, anxiety, compassion and empathy and during the birthing process.

These are just three regulatory constituents of our signaling pathway. There are numerous others. These regulatory biochemicals may vary to some degree between individuals. The activity level of your neurotransmitters may not have the same activity level as other individuals may have. Neurotransmitters, neuronal structures and the associated biochemicals may exist in several variant forms, from one human to another, the result of slight differences in one's DNA. One small chemical difference may have profound consequences. These variants may be responsible for some of humans more bizarre behaviors.

Deoxyribonucleic acid, DNA, is the molecule that encodes the genetic instructions used in the development and functioning of most all known living organisms; it is essentially the blueprint for that organism. DNA is a very long double-stranded molecule. Each strand is composed of a chemical backbone and, in a specific sequence, pairs of four nucleotides; adenine, thymine, guanine and cytosine and abbreviated *A, T G* and *C.* The running sequence of four nucleotides is the 'alphabet' of gene coding. These specific sequences are essentially the recipe for a protein. If a mutation should happen to occur during the replication of a long DNA sequence, for example, a *T* is replaced incorrectly with a *C,* the sequence would then be slightly altered. This small mutation in the DNA could alter the "recipe" such that the newly synthesized protein may still be functional, or it may not function, or it may have an altered function. A mutation may alter one base pair or the base pair could be deleted; this latter occurrence is called a deletion. Mistakes in the replication of DNA occur randomly and rarely.

As an example, the normal function for an enzyme called catechol-O-methyltransferase, or COMT, is to maintain appropriate levels of certain neurotransmitters, such as dopamine and noradrenaline in the prefrontal cortex of the brain. The prefrontal cortex is involved with personality, planning, inhibition, abstract thinking, emotion, and short-term memory. A mutation in the COMT gene, actually a deletion, may cause an individual to develop schizophrenia, depression, anxiety and bipolar disorder. May factors have roles in complex disorders such as schizophrenia and small mutations in the DNA may be one of the factors. [10]

# Four

## Genetic Behavioral Traits

*Nature is all that a man brings with himself into the world; nurture is every influence that affects him after his birth.*

—Francis Galton (1822-1911), the English Victorian polymath, first coined the phrase *Nature versus Nurture* as used in its modern sense. Galton was a cousin of Charles Darwin. Two centuries earlier, the viewpoint that humans acquire all or almost all their behavioral traits from nurture was termed *tabula rasa* (blank slate) by John Locke (1632-1704), an English philosopher and physician.

Bacteria could be thought of as having behavioral traits. They move by their flagella, microscopic whiplike appendages that enables many protozoa, bacteria, spermatozoa, etc., to swim along a food gradient. They obviously *notice* which direction to navigate, towards the higher concentration. And, once at the source, they turn on a new set of genes that will assist in bringing the nutrients into the cell. Then other genes will be switched on to synthesize enzymes that can transform the food into energy. So do bacteria have behavior? A sunflower blossom follows the path of the sun. Is this a behavior? Obviously these traits are genetically determined; a consequence of their DNA.

Plants respond to signals in their environment. Gravity and sunlight *cause* plants to send their roots down and the stem upwards. Some plants wage chemical warfare against their pests and annoyances. Plants have developed unique tactics for survival.

When I was a botany student at UCSB we went several three day field trips. One such was Dr. Bob Haller's taxonomy field trip to the Mojave desert to study desert vegetation. One plant we studied had some very with interesting characteristics; *Larrea tridentata,* commonly called the Creosote Bush. This plant may be unfamiliar to most but it is rather common in the southwestern deserts. This a most unusual plant. The whole plant exhibits a characteristic odor of creosote from which the common name derives. It's an evergreen shrub growing to 1 to 3 meters with small leaves and yellow flowers. It is not particularly beautiful but its traits are very interesting. It flourishes rather well in an extremely harsh desert environment. The leaves of the creosote bush are coated with Nordihydroguaiaretic acid, a resinous phenolic compound which slows evaporation and protects the plant from being eaten by most mammals and insects. Once leaves are ingested, the plant's resinous compounds form complexes with the animal's proteins and prevent digestion of the leaf tissues. Jackrabbits and wood rats are the only mammals known to eat the leaves of this plant and then only in times when little else is available.

There is another trait of the creosote bush that is also interesting. In its desert environment, where water is scarce, the phenolic resins produced by these plants inhibit the growth of nearby plants producing areas of little or no growth of other species, commonly called fairy rings.

Likely unknown to most is that the creosote bush one of the oldest living organism on Earth. Some individual bushes have lived an estimated 11,700 years. When a plant becomes 30 to 90 years of age, the central crown dies back and splits into separate viable crowns. These separate crowns all originated from one original seed. A single clonal colony of the creosote bush may spread out to 14 to 20 meters in diameter. Older than Redwood trees, older that the Bristlecone pines, this odoriferous desert plant hold the record on longevity. It is truly the supreme survivor of the plant kingdom.

Most people would conclude that while plants have some interesting and complex characteristics, they certainly do not have Free Will. The concept of Free Will has been debated for centuries and there is no universally accepted definition. Most humans assume that we do have Free Will. Humans, in a fashion, are not too dissimilar from plants. We too respond to environmental signals in almost a predictable manner. We have developed unique traits that have ensured our survival. Yet we feel that we have Free Will, something special that separates us humans from the rest of the animal kingdom. Maybe we should not rush to any conclusion about whether we do or don't have Free Will. As some human traits are presented, it might be interesting and worthy of further thought, the concept of Free Will.

There are some 200 human *physical* traits that most agree are genetically determined. Handedness, free or attached earlobes, eye color, hair color, straight or curly hair, sexual orientation, allergies, schizophrenia and colorblindness are just a few of the physical genetic traits.

Humans have numerous behavioral traits such as love, compassion, fear, anger, rage and our tendency to gather in groups that have cultural similarities. We know that some of our human behavioral traits are inherited genetically from our ancestors DNA and that some behaviors are cultural, that is learned, traits passed down from one generation to the next through a sort of schooling. Obviously the way people behave is a complex interaction between a great number of genetic, social and environmental factors. Genes determine the general structure and biochemicals of the brain. Genes also determine how the brain reacts to our experiences. Experience is required to enrich the matrix of synaptic connections. Individuals raised in an environment of higher levels of stimulation, especially during critical periods of development, have increased numbers of synapses with greater complexity. Youngsters raised in a sensory deprived environment, such as in some old Soviet-style orphanages, have delayed and impaired cognitive development. Individuals with higher education who participate in cognitively stimulating activities have a greater resilience to the effects of aging and dementia. The nurturing component of human behavior is critical and well studied. It is the genetic component that, in part, constitutes the nature of humans.

Humans evolved with traits appropriate to be the last hominin standing. All the constituents of our brain either function properly or they misfunction, resulting in the myriad of bizarre genetic behaviors exhibited by many humans. The scientific literature on genetic behavioral traits is enormous. We will explore some examples from the scientific literature which will be simplified. The brain's biochemistry, along with variant pathways and biochemicals, will illustrate that our behavior is the result of our brain activity. Duly noted is that the expression of any behavior is dependent our genetics, prior learning and experiences, as well as situational and environmental factors.

For clarity, *Trait* will be used as a distinguishing quality or characteristic that presumably has genetic components; it is inheritable. So as not to be misleading, all the factors that comprise the biochemicals and neuro-pathways any specific trait may not be completely understood. Even simple traits such as love may actually be composed of numerous biological pathways that could, with deeper understanding, be several separate traits. Presented are a few humans traits identified with names that we can easily associate. It will not be a comprehensive review of all traits.

We all start as a fertilized egg containing DNA from each of our parents, and here we are, mature adults with amazing behaviors. The pathway from our DNA to our social behavior is complex and convoluted. Genes are expressed either singularly or in concert with other genes. Genes are the code for the proteins and enzymes of the brain, the neurological networks, transmitters, transporter and receptors. With all these constituents together with our memories, acquired knowledge and the environment, we find ourselves behaving as human beings.

# Trait: *Fear*

The New Yorker, July 29, 2013, displayed a cartoon by Roz Chast that seems appropriate for the subject of fear. The cartoon depicts the cover of a fictional magazine, one you might find on the newsstand alongside the tabloids displayed at the supermarket checkout. The fictional magazine is titled *New Dread, the Magazine of Undiscovered Fears.* Featured on the cover are the titles of articles within the magazine: *The Ten Reasons to Avoid Bananas, Should Sofas Be Banned? The Case against Flannel* and *Is Your Tea Putting You at Risk?* Knowing some humans, this magazine might be quite appealing.

Fear is the most simple and evolutionally conserved type of memory in mammals. From lizards to humans, the key area of the brain responsible for memories of fear is the amygdala. Fear and anxiety are also linked. Fear just might be the most prevalent trait among humans. It is a modulator of human behavior. We face fear daily, sometimes a real danger while driving. Parents are fearful for the safety of their young children. At times fear is used by politicians as a method to sway the public to their view; for example, the enemy is close at hand, let's invade. These politicians, brave souls, present apparently fearful scenarios followed by solutions that only they can provide.

You most likely are aware that some individuals have greater fear than others and some people seem not to have great fear. Little fear is common among young males. Not everyone has the same degree of fear for the same situation. Several factors are involved in the degree of fear one has. Certainly prior experiences are important. Experienced airline pilots have little fear when landing a plane. This would not be true for an untrained individual. But there is an important factor to consider. Fear has a genetic component, and this may explain why some people have great fear that others may not have. The level of one's fear is, in part, dependent upon your genetics. Fear can overwhelm rationality. Fear often leads to exaggeration, suspiciousness, conspiratorial fantasy, and eventually to hysteria. Politicians realize this. Planting fear is their path to power.

Math anxiety is common among many people. Investigators performed a series of brain scans on second- and third-grade students while they did addition and subtraction. They discovered that those who feel panicky about doing math had increased activity in the amygdala, the brain's main fear center. There was also increased activity in a section of the hippocampus, a brain structure that is involved in problem-solving and new memory formation. The results suggest that, in math anxiety, math-specific fear interferes with the brain's information-processing capacity and its ability to reason through a math problem. [1]

### *Stathmin and Fear*

Scientists have found a gene for fear. This gene codes for the protein called Stathmin, which is involved in nerve-cell communication and is required to form fear-related memories. Normally, when mice are placed in new surroundings, they naturally avoid open or exposed spaces. Scientists have genetically manipulated the mouse's genome to remove the Stathmin gene, a procedure call *knock-out* technology. These Stathmin knock-out mice do not seem to have instinctive fear. They explore open and unfamiliar spaces, which normal mice would avoid. These mice without the Stathmin gene would not survive for long out in the field. Stathmin-knock-out mice exhibit a decreased memory in amygdala-dependent fear conditioning and fail to recognize danger in innately aversive environments. These knockout mice have reduced ability to form fear-related memories. [2]

Stathmin

A gene that is expressed in the Amygdala
No Stathmin gene (in mice) = no fear.

### *Serotonin and Fear*

The level of fear one has is extremely more complicated than just this one protein. The amygdala is usually involved, as well as the brain's neural network, neurotransmitters and the factors that regulate the level of these chemicals. [3]

Serotonin, a neurotransmitter, is associated with fear as well as with cognitive functions including memory, learning as well as our feelings of well-being and happiness. It functions as a regulator of mood, appetite, and sleep. Reduced levels of serotonin is associated with fear, depression, anxiety and aggression.

Serotonin levels in the brain are controlled by several factors, one of which is the Serotonin Transporter gene that moves serotonin along the neural network. Some individuals have a shortened version of this gene, which results in lowered serotonin transport and consequently lower serotonin levels. Individuals who carry this shorter transporter gene have a higher level of anxiety and fear than individuals with the longer version. [4]

Serotonin and Fear

Serotonin Transporter gene - Two versions
Short version = low serotonin = anxiety and fear
Long version = normal level and normal behavior

Another gene that influences behavior via serotonin levels is the *Serotonin Receptor gene*. The product of this gene is the *serotonin receptor*, the protein which binds serotonin, regulating the flow of ions across the synaptic cleft. In laboratory mice it has been possible to manipulate the expression of this gene. If the Serotonin Receptor is expressed at high levels, the firing of serotonin neurons is reduced commensurate with higher aggression. [5]

Serotonin Receptor Protein binds Serotonin which lowers Serotonin levels.

High Levels of the Receptor Protein = Lower levels of free serotonin = reduced neuron firing = Higher Aggression, Psychotic disorders, Violence, High Sex Behavior.

Low Levels of the Receptor Protein = Higher levels of Serotonin = Normal Behavior.

The serotonin receptors are the target of a variety of pharmaceutical drugs, including many antidepressants, antipsychotics, antimigraine agents as well as hallucinogens. Patients with excessive fear have been successfully treated with the antipsychotic drug lurasidone. This drug modulates the serotonin transporter gene activity. [6]

Humans have a fearful nature. We fear taxes. We fear restrictions on our behavior. We fear we will not have enough. We fear poisonous snakes. We fear the oppositional political party. Fear influences far too many of our decisions. We fear the unknown, and our reliance on religion for answers is far too common. And we rely upon those scary TV political ads for guidance.

Politicians send messages of fear: *The illegals are rapists and they are streaming across our borders.* They want your vote. But without fear we most likely would have perished.

When my daughter was growing into a young woman, about age twelve, I had an opportunity to teach her how to deal with certain fears. We were walking in a rather seedy section of San Francisco after dark. A young man wearing a hoodie, baggy jeans and white sneakers with the laces untied was approaching us. He appeared a bit suspect. I remember these details because I followed the advice I gave my daughter. Before he was too close to us I told her to not look down at the pavement, don't exhibit fear but instead look him over top to bottom and take a mental note of his physical appearance including his shoes and clothing. Just look him over in an investigative fashion. Don't smile, don't look aggressive, but do, respectfully, acknowledge him with a slight nod. I do not know if he had any ulterior agendum other that to get to his destination but the fact that we could have identified him may have prevented an unwelcome event. He just kept on his way. One's fear telegraphs a signal. You have something of value. on, anxiety and bipolar disorder. Many factors have roles in complex disorders, such as schizophrenia, and small mutations in the DNA may be one of the factors. [10]

## Trait: *Anger and Aggression*

*Anger dwells only in the bosom of fools.*

—Albert Einstein (1879-1955)

For centuries, it has been observed that with many birds, dogs, fish and mice, aggressive behavior can be selected and bred. In other words, aggression has a genetic component. The genes we inherited from our ancestors determine the degree of anger we exhibit when faced with a triggering situation. Some individuals easily provoked and lash out with anger and aggression. A confrontation may result from misunderstanding, miscommunication or selective or false memories. For most individuals this behavior is tempered by one's parenting or the presence of a police officer.

A few teenagers often make headlines in the newspapers and not with the stories we had hoped. These children in their early years were cute, well-mannered and adorable. Unfortunately a few of these cherubs turn into callous and unemotional terrors in their teen years. What were they thinking? Turns out that it might not only be the parenting they received; for some it's their genetics.

In youths, there is a correlation between aggression, callous and unemotional behavior and amygdala activity. Studies conducted by Dr. Abigail Marsh and her colleagues at Georgetown University consisted of 46 male and female youths aged 10 to 17. 30 of the youths were judged as having callous-unemotional (CU) trait issues. The control group consisted of 16 healthy participants. Functional Magnetic Resonance Imaging (fMRI) visualizes which parts of the brain have intense electrical firing of nerve circuits and intense blood flow and oxygen consumption. fMRI analysis found that the amygdala's response to fear was significantly lower in the teenagers exhibiting the CU trait. Events that would be perceived as fearful in a normal teenager would not elicit a fear response in youths that were callous, unemotional and aggressive. Youths that did respond to a fearful cue were more empathetic and less aggressive. [7]

Anger and aggression are not a one gene-one behavior phenomenon. There are numerous event factors that take place in the brain that influence levels of anger and aggression. Serotonin, dopamine, oxytocin, MAO-A, DARPP-32, testosterone, the hypothalamus, the amygdala and a whole bevy of the brain chemicals and structures are intertwined, and manifested as anger and aggression. And as you can imagine, with so many components, thing do go astray. Following are a few examples that illustrate the science as to why there are differences is levels of anger and aggression between individuals.

Monoamine oxidase A (MAO-A) is an enzyme that degrades some neurotransmitters, such as dopamine, norepinephrine, and serotonin. The gene as well is referred to as MAO-A. Researchers at Barcelona University have demonstrated a correlation between the *promoter*, or switch, for the MAO-A gene and aggressive behavior. This promoter is not identical in all humans; there are polymorphisms or variants in MAO-A gene's promoter. [8]

The aggressiveness of 57 university students was measured by means of the standard *Aggression_Questionnaire.* This was compared by genetic testing to which particular variant of the MAO-A promoter the students carried. Individuals with one particular promoter variant, which causes lower expression of the MAO-A gene, were more likely to display aggressive or anti-social behavior when placed in an environmental adversity situation than individuals with the normal MAO-A promoter. [9]

MAO-A regulates dopamine, norepinephrine and serotonin
normal MAO-A promoter = normal MAO-A expression —> normal behavior
variant MAO-A promoter = low MAO-A expression —> aggression, anti-social

It has been shown that when levels of the neurotransmitter dopamine are elevated, test animals are quicker to display anger and aggression.

Another factor in the level of anger and aggression expression involves the regulator protein DARPP-32 (Dopamine and cAMP regulated neuronal phosphoprotein) is a protein that regulates the level of dopamine. The gene, which codes for DARPP-32 exists in three naturally occurring variations amongst humans.

Researchers at the University of Bonn gave a questionnaire to 838 individuals to gauge their anger temperament, how they responded to stressful situations. They also sampled the test participants DNA to determine which of three versions of the gene coding for DARPP-32 they were carrying.

The distinction between these three variants of the DARPP-32 gene is rather small, but significant. At one specific point in the DNA for this gene, if the sequence of nucleotides is thymine-thymine (T-T) or thymine-cytosine (T-C), these individuals were significantly more angry and higher levels of dopamine than those with the Cytosine-Cytocine (C-C) version and lower levels of dopamine.

This very small naturally occurring genetic variation in the DARPP-32 gene found across human populations may have a direct affect on human behavior. DARPP-32 is one key to understanding why some people display the personality trait of greater anger. In addition, fMRI data supported the role of the amygdala for the processing of anger. [10]

Rewards, food, and sex increase levels of dopamine. Addictive drugs increase dopamine neuronal activity.
Gene for DARPP-32 has three variants.
T-T and T-C = Heighten levels of Dopamine = Higher Anger
C-C = Lower levels of Dopamine = Normal Anger

GABA (γ-Aminobutyric acid), is a small molecule that is the chief inhibitory neurotransmitter in the mammalian central nervous system. Its activity is associated with states of anxiety, antisocial and addictive behaviors. GABA plays the principle role in reducing neuronal excitability throughout the central nervous system. GABA is also found in plants. It is the most abundant amino acid in tomatoes. It may have a role in cell-signaling in plants.

The reasons why some men act impulsively, rather than rationally, may be related to a lower concentration of GABA receptors in the brain. Men with higher levels of GABA in the prefrontal brain region have a tendency to behave calmly. Men with lower GABA levels are more likely to act aggressively, drink and take drugs in response to distress or strong emotions and urges. Interestingly, GABA is a negative regulator of dopamine function. [11,12]

Men with low levels GABA + distress —> impulsive, aggressive, antisocial and addictive behaviors.
Men with higher levels GABA + distress —> behave calmly.

Testosterone dynamics also influence aggression. In monogamous birds, testosterone levels rise at puberty to moderate levels. Sexual arousal and challenges involving young males increase testosterone levels further. When males are required to care for offspring, testosterone levels will decrease. In human males, testosterone levels are also associated with different behavioral profiles, associated with mating or parental efforts. [13,14]

Review:

Teenagers with low amygdala expression may be less empathetic and more aggressive.
Genetic variants of the MAO-A *promoter* results in low MAO-A expression, which may induce aggression.
DARPP-32 gene variants may increase anger response.
Low GABA levels have been linked to aggression.
Testosterone levels influence aggression.

Before we move on, here are some thoughts concerning anger and aggression. I would think that most all of you may know someone who is excessively angry and aggressive. Do they lack self-control or are there some over-riding genetic factors involved? If a genetic analysis were readily available which would determine if an individual carried any of the variant "anger" genes, have you any suggestions as to which individuals should be assayed? One particular former politician come to my mind. Perhaps all politicians should be assayed prior to elections. Perhaps the best test of behavior is to just judge how anxious they are to go to war.

## Trait: *Rage and Revenge*

*Revenge is often like biting a dog because the dog bit you.*

—Austin O'Malley - a Gaelic footballer

Family feuds, dysfunctional families, brother versus brother, adult children despising a parent, family fights that last decades, sometimes the duration of their lives, even generational intra-family battles. What is going on? Individuals displaying these behaviors are accusatory, exhibit warped memories regarding relevant events and react in a vindictive manner. They won't consider alternative explanations, and logic and reason are beyond their grasp. Rage increases as a function of time.

Revenge is certainly prevalent among some humans. Many people are willing to incur costs to punish others who, in their estimation, have violated some social norm. Onset of these episodes may be initially caused by our inability to rationally think of all possible alternative explanations for the slight we received. Miscommunication or confusion of words or signals, faulty or selective memories, or insufficient information is often involved. Consequences of these actions are rarely considered. Most people know of such occurrences; they are not that rare. When encountering such a situation, it may be better to just walk away. Change may not be possible. These individuals may have genotypes that predispose them to such behavior.

*This is certain, that a man that studieth revenge keeps his wounds green, which otherwise would heal and do well.*

—Francis Bacon (1561-1626), an English philosopher, statesman, scientist, jurist, orator, essayist and author. Bacon has been called the father of empiricism.

Unreasonable behavior, such as rage or revenge, may be caused by biochemical or anatomical "imbalance" in the brain. There are a plethora of personality disorders that are recognized in The Diagnostic and Statistical Manual of Mental Disorders manual published by the *American Psychiatric Association.* The manual describes disorders that cause people to have limited capabilities in their ability to function normally. One such disorder is *Oppositional Defiant Disorder*, a childhood disorder in which there is an ongoing pattern of anger-guided disobedience, hostility, and defiant behavior toward authority figures that goes beyond the bounds of normal childhood behavior. Children suffering from this disorder may display stubbornness, anger, aggression and antisocial behavior.

There is an Eastern European folktale in which a genie offers to grant a man's wish as long as his hated neighbor gets double the prize. The man says,

*"Put out one of my eyes."* Why is it that some individuals would punish, humiliate, even torture others even when they themselves suffer a cost?

The *Spitefulness Scale* is an assay used to determine the magnitude of spitefulness. A group of scientists tested a large sample of 946 college students and cross-validated with a national sample of 297 adults. Test participants were asked to rate how firmly they agreed with sentiments such as: "If my neighbor complained about the appearance of my front yard, I would be tempted to make it look worse, just to annoy him, or If I opposed the election of an official, I would happily see the person fail even if that failure hurt my community, or I would be willing to take a punch if it meant someone I did not like would receive two punches." Men reported higher levels of spitefulness than women, younger people were more spiteful than older people, and ethnic minority members reported higher levels of spitefulness than ethnic majority members. Spitefulness was positively associated with aggression, psychopathy, Machiavellianism, narcissism and low guilt. Spitefulness correlated negatively with self-esteem, guilt-proneness, agreeableness and conscientiousness. [15]

Agreeable, conscientious people with self-esteem = not spiteful

What is gained by the individual who seeks revenge? Scientists have discovered that subjects who seek revenge actually derive a satisfaction reward. During revenge episodes, the *dorsal striatum*, a subcortical part of the forebrain, is activated in the same manner as in the processing of rewards, novel, unexpected, intense or aversive stimuli. [16]

## Trait: *Delusion*

*It is far better to grasp the universe as it really is than to persist in delusion, however satisfying and reassuring.*

—Carl Sagan

Humans are susceptible to delusions that range from the mild, *"I'm sure my horse will win the race"* to the severe, *"I'm always being followed."* The clinically recognized *Delusion Disorder* is where the patient suffers from persistent delusions that could not possibly be true and the symptoms persist for at least one month. The non-bizarre delusions typically are beliefs of something occurring in a person's life that is not out of the realm of possibility, yet has no veracity. The Delusion Disorder is not associated with the effects of a drug, medication, medical condition or schizophrenia. A person with Delusional Disorder may be high-functioning in daily life since this disorder bears no relation to one's IQ. Examples would include conspiracy theories, spousal cheating, believes someone will harm you or is spying on you. Fact-checking can dispel these beliefs for the rational. The human brain has evolved to have an enhanced threat detection system. The intensity to which we respond is dependent upon the degree of sensitivity we have to a threat and whether we fact-check or look for additional evidence. If that system becomes oversensitive, the result is paranoia. One's physical and chemical makeup and of course our environmental experiences may determine how we respond. Political operatives may just be using our susceptibility to delusion to advance their cause. Just a thought.

One treatment for delusional patients is the prescription of an antipsychotic medication called Haloperidol. This drug functions as an *inverse* agonist of dopamine, meaning that the binding of the drug *decreases* the activity, in this case dopamine, to a lower basal level. This drug is a dopamine-blocking agent. [17]

Delusional patient = high dopamine —> paranoid symptoms
Delusional patient + antipsychotic drug = decreased dopamine —> normal

*Self-delusion, in other words, feels good, and that's what motivates people to vehemently defend obvious falsehoods.*

—Dan Kahan, Professor of law at Yale Law School. [18]

The psychological theorizing of emotional regulation, anxiety in particular, was of interest to Sigmund Freud, who made anxiety-regulation one of the centerpieces of the psychodynamic theory of mental life. [19]

*It is not to be forgotten that what we call rational grounds for our beliefs are often extremely irrational attempts to justify our instincts. In matters of the intellect, follow your reason as far as it will take you, without regard to any other consideration... In matters of the intellect, do not pretend that conclusions are certain which are not demonstrated or demonstrable.*

—Thomas Huxley (1825-1895), an English biologist, known as "Darwin's Bulldog" for his advocacy of Charles Darwin's theory of evolution.

Creationists, who believe that God created the universe 10,000 years ago, believe that ancient dinosaur fossils were planted in various strata by God to test the faith of humans. Bertrand Russell proposed, as a counter to this irrationality, that perhaps God created the universe 20 minutes ago, just as it is today, with buried fossils as well as luxury cruise ships, skyscrapers and football stadiums. God also provided everyone with the memory they now have. Bertrand Russell's proposal would be impossible to disprove. William of Ockham (c. 1287-1347) would be frowning at such a proposal.

Ockham's razor, attributed to William of Ockham, means *law of parsimony.* Fittingly it was Bertrand Russell who interpreted Ockham's razor as "if one can explain a phenomenon without assuming this or that hypothetical entity, there is no ground for assuming it, i.e., that one should always opt for an explanation in terms of the fewest possible causes, factors, or variables." [20]

## Trait: *Motivated Reasoning*

*It is the peculiar and perpetual error of the human understanding to be more moved and excited by affirmatives than by negatives.*

—Francis Bacon

Motivated reasoning is, in my opinion, a dominant factor in what makes humans contentious. It seems inevitable that whenever a group of individuals are gathered around chatting and the topic of religion or politics is brought up, one can be sure that it can lead to a contentious situation if there are differing views. If an opinion is offered which differs from others, sides are taken and disharmony arises. Rational debate is rare. Arguments ensue, with each side offering arguments and objections and rarely addressing the other's concerns. *Why does your candidate condone strict voter laws?* Rather than addressing the question put forth, the other person will offer in rebuttal another question such as *Why does your candidate support gun registration?*

These debates do not involve simple reasoning, such as finding the length of the hypothesis of a right triangle knowing the lengths of the two adjacent sides. It is more complex.

Confirmation bias refers to a type of selective thinking whereby one tends to notice and to look for what confirms one's beliefs and to ignore or undervalue the relevance of what contradicts one's beliefs. Conformational bias is an *explicit,* regulated, emotional behavior in that it requires conscious effort for initiation and some level of monitoring during implementation. It is associated with some level of insight and awareness.

Motivated reasoning is an elevated form of confirmation bias in which *implicit* or automatic emotion regulation in the brain converges on judgments that maximize the positive and minimize the negative. Implicit processes are evoked automatically by the stimulus itself and run to completion without monitoring. It happens without insight or awareness. Motivated reasoning is a coping mechanism when faced with oppositional facts. [21] Our beliefs motivate us to accept new information that is allied with our beliefs and to reject alternative information. People may develop elaborate rationalizations to justify holding beliefs that logic and evidence have shown to be wrong. When confronted with information contrary to firmly held beliefs, the brain is able to minimize the negative and maximize the positive in a automatic way. Contrary information motivates us to reject such without discussion or contemplation. Humans often respond with reasons that favor their views. [22]

There have been numerous studies of motivated reasoning. One such was performed during the US Presidential election of 2004. Committed partisans, when involved in political judgments of their candidate or the opposition, showed activity in their brain associated with motivated reasoning or implicit reasoning and little neural activity in regions associated with logic-based or explicit reasoning. [23]

There is a compulsion to justify decisions, even if they may be irrational. Once a decision has been made, second-guessing our previous decision locks in the new decision and prevents us from over- pondering that decision. Tested humans, as well as some monkeys, seem to derogate alternatives they have chosen against, changing their current attitudes and preferences to more closely match the choices they made in previous decisions.

A perfect example is in Aesop's fable about the fox who stopped trying for the grapes, with the quip *they were probably sour anyway.* Quipping "They're probably sour anyway" makes us feel better. Our initial choice, the grapes, being unavailable, the alternative option of walking away is just fine. So in one sense, don't blame a person for being irrational and unreasonable. They can't help it. They are a prime example of evolution. Unfortunately, most humans will not change, see the light, be logical and be reasonable. That may be our nature. [24]

Nearly everyone has had an encounter with a person obsessed by conspiracies. I have an acquaintance who believes that this earth is not warming. Additionally he believes that the scientific community is conspiring for some liberal agenda. He is familiar with my concerns about anthropomorphic global warming. I have been following the scientific literature for several decades. Unexpectedly this individual emailed articles that had been posted on anti-climate change blog sites, articles that had not been peer-reviewed. The articles sent had misrepresented facts and made conclusions not based upon scientific evidence. I assumed that he would like the accepted scientific perspective. I responded by providing reasoned scientific explanations. I included questions and concerns in my return email. The response never addressed my concerns nor answered the questions I ask. His response was to send additional non-peer reviewed anti-climate change articles. One email I received: *We don't need thousands of scientists, but only one that provides evidence or an accurate scientific hypothesis.* My response, *If one person states a hypothesis and 10,000 other respected scientists reject that hypothesis, would it not give you pause to reconsider the first person's hypothesis?* Instead of responding to my question, his response was, *Why, if global warming is such a crisis, it ranks near the bottom of the list of priorities Americans have?* I responded that the anti-global warming campaign by the fossil fuel industry has been very successful. Again no response to my questions, just additional emails with more anti-climate change blogging. There is an abundance of scientific research papers. He refuses to read these papers; instead he relies upon blogs that attempt to discredit the research. All this was very annoying. I just stopped responding. His reasoning is motivated. He will never change.

## Trait: *Selfishness and Greed*

In ancient tribal times, *Homo sapiens* needed to be selfish to insure sufficient assets to survive. Greed would be excessive selfishness, obtaining assets far in excess of one's needs for survival. There are those who believe that some is good, more is better and too much is just right. The greater resources one can accumulate, the greater chance that their genes will survive into the next generation. Sharing one's wealth is not always a beneficial trait.

Obtaining resources such as food or mates may often involve direct conflict. Aggressive individuals may have an edge in obtaining resources. A cunning and aggressive nature would allow organisms possessing these traits to out-compete other organisms for scarce resources, and hence the genes of such an individual will increase in frequency within the population.

Selfishness, in the sense of self-survival, is a self-preservation trait that is exhibited when one's life is in jeopardy. Individual survival instincts, while not completely understood, can be observed from microbes to primates. An organism either recognizes a danger and acts responsibly or risks demise. Bacteria will move away from a toxic environment. Plants have various physical traits that ensure survival. Spines or toxins deter predators. In dangerous situations some animals freeze, some flee, some attack. Forming a herd is a survival tactic, safety in numbers.

Selfishness has a genetic link. Researchers at the Hebrew University in Jerusalem have identified a presumptive *selfish gene.* They have proposed a link between a gene called *Arginine Vasopressin Receptor 1A (AVPR1A)* and ruthless and selfish behavior. The AVPR1A gene is a possible candidate for selfishness. It is known to code for receptors in the brain that detect vasopressin, a hormone involved in altruism and prosocial behavior. The AVPR1A gene is polymorphic; it codes for two different alleles, the short and long versions.

The study at the Hebrew University consisted of 200 students whose DNA was tested for which variant of the *AVPR1A* gene they processed. The game, dubbed *The Dictator Game,* consisted of opponents sharing or not sharing wealth, behaving greedily or behaving generously. There was a positive correlation between the length of the *AVPR1A* gene and greediness. The study showed that people were more likely to behave selfishly if they had the shorter version of this gene, concluding that the reward centers in the brain of the greedy ones may derive less pleasure from altruistic acts and hence may act in a more selfish fashion, not sharing their wealth. [25]

AVPR1A : Arginine Vasopressin Receptor - A Protein that Binds Vasopressin.

There are two Variants in the DNA sequence that codes AVPR-1A:

Long Version (—————————) Generosity, Altruism, Pleasure
Short Version (——————) Selfishness, Greediness, less pleasure from Altruism

In the 1976 book The Selfish Gene, Richard Dawkins uses the term *selfish gene* as a metaphor, a way of expressing the gene-centered view of evolution as opposed to the views focused on the organism and the group. From the gene-centered view, the more two individuals are genetically related, the more likely selfless behavior will spread through the population. No *selfish gene*, as such, was described.

An actual gene for a selfish behavior, a gene that would presuppose one to be selfish may not exist as a separate gene. Selfishness most likely arises by a suite of genetic factors, brain components and structures, including variations of the AVPR1 gene.

Humans can be selfish, wanting immediate results regardless of long-term consequences. We may steal goods from others, swindle, connive and be opportunists. We may only consider our needs at the expense of others. Sometimes some of us respond with anger and aggression. But without attending to one's own needs, we would not have survived.

## Trait: *Stubbornness*

*I'm not stubborn. My way is just better.*

—Maya Banks, American author of erotic romance novels.

We all seem to know people who are stubborn. Some forge ahead without sufficient caution or knowledge of the consequences. David Farragut, a Flag Officer during the American Civil War, was such a stubborn person. On August 5, 1864 during his siege of Mobile Bay, after one of his ships sank in the heavily mined bay and the rest of his fleet was pulling back, Farragut gave the famous order: *Damn the torpedoes, full steam ahead.* The bulk of the fleet entered the bay and were triumphant. Tethered naval mines at the time were known as torpedoes.

Some people forge ahead without fully considering all the possible consequences, hoping for a favorable outcome. Obviously, knowledge of consequences and a bit of luck favor a positive outcome. Entrepreneurs who risk their time on unproven ideas may have a bit of a stubborn streak. Calculated risk may result in a positive outcome unrealizable without taking the chance.

There are some twin studies that suggest that stubbornness may have a genetic link. However, the correlation of these traits in identical twins as compared to non-identical siblings is just slightly positive. [26] As of yet there are no definitive studies that would indicate that stubbornness is a genetic trait. It most likely involves a suite of genetic traits along with environmental considerations.

## Trait: *Denial*

*It's not denial. I'm just selective about the reality I accept.*

—Bill Patterson (b. 1958), creator of *Calvin and Hobbes* from 1985 to 1995.

This is a question I have pondered: do Sean Hannity and Rush Limbaugh, conservative media personalities, truly believe that the climate crisis is a hoax? I suspect that these two individuals are playing with their conservative base in the manner of shysters who sell snake oil to naive and gullible people, knowing full well that it is useless. Huge audiences equate to big money. Are they deniers with economic objectives because to behave otherwise might decrease their income? Perhaps they are deniers in a clinical sense, deniers in the same manner as Holocaust deniers and some alcoholics? We may never know.

Denial is a self-preservation mechanism, a coping mechanism that allows one time to adjust to or minimize a painful or stressful issue. Denial is an unconscious process where the brain prohibits perception. There is an interesting book by Ajit Varki and Danny Brower. They argue that denial is a prerequisite for human intellectual development. Empathy is the awareness of another person that can only occur when, evolutionarily, we are aware of ourselves. *Denial of reality is an essential skill for us to function normally in the world. It is a fundamental property of being human.* [27]

## Trait: *Prejudice*

*Prejudice is a great time saver. You can form opinions without having to get the facts.*

—E.B. White (1899-1985), an American writer.

When I was about 7 or 8 years of age, our family lived in Monterey, California. This would have been near the end of WWII. Monterey at the time had a robust sardine fishing industry. Many of the fishermen were Italian and Catholic. My father was a bit of a bigot. He didn't care for Catholics or Italians. I never knew why. I remember him saying that Catholics ate fish on Fridays. I don't ever remember him eating fish on any Friday. I think he didn't want to be identified as a Catholic. I also remember that he also didn't care for Jews. He mentioned that our neighbors just down the street were Jews, stated in a manner that indicated his disdain of Jews. No reason for his dislike was given. On a day shortly after he made that statement, I was playing with the son of this Jewish family. I don't remember the boy's name, but I do recall that he was about a year older and a bit larger. We must have had some sort of disagreement and he gave me a big shove. I knew that pushing him back would only escalate the situation with me ending up the loser. I immediately said, *You're a Jew*. He drew back *as if* I had landed the greatest of punches. Not another assault occurred. He just went home. My thought at the time was that calling him a Jew was powerful. I caused more damage than if I had actually punched him. I didn't see him or play with him after that. This incident caused me to wonder just what a Jew was. I had no idea. The word Jew was just a word I didn't know. I was completely ignorant. When one intentionally denigrates another, is it simply a matter of bolstering our self-esteem? Is there a genetic factor, a primitive instinct involved with this behavior? If there is a genetic component, what would be its survival value? Or is prejudice purely a consequence of our nurturing?

Prejudice remains as *vibrant* today as it was decades ago. When members of another tribe reside outside of our circumscribed boundaries, some individuals resort to various forms of discrimination. We are cautious of anyone unlike ourselves. Race discrimination remains. Until the mid-twentieth century the word nigger was used to identify members of the black community. I recall my grandmother using that word when referring to a black person. My grandmother was a warm, kind and caring individual. She made the best cinnamon rolls ever. But she came from Wisconsin, a state where she was not exposed to the black community to any extent. She must have learned that word as a child and, knowing no difference, continued to use the word for any black person regardless of the circumstances. I'm sure she did not want to disrespect these people.

There is a qualitative difference between prejudice and bias. Prejudice refers to pre- judging, making a judgment without consideration of all facts. Having a bias is common with all people. Some people prefer chocolate ice cream over vanilla. They have a bias. It is doubtful that these people who prefer chocolate ice cream have contempt of those with different preferences. With respect to my grandmother, it gets a bit sticky. Had she a bias or was she prejudice?

When President Barack Obama became a candidate for President, he came under intense scrutiny. Conservatives questioned his birthplace and his religion. One individual told me just prior to the election that if Obama were elected, the White House garden would be planted with watermelons. Few would dared called him a nigger, since it is now, thankfully, a disrespected word. Progressive ideals advance slowly.

*Born this way.*

—Lady Gaga

What great insight! Why is it that some humans are homosexual? Can homosexuals be converted to heterosexuals? Can heterosexuals switch over to be homosexual? Why are some humans homophobic? To me, as a straight male, the idea of having a sexual encounter with another male is quite repulsive, to the point that I don't even want to know or think about what two males do. Why is that? No one told me to think of it as repulsive. I didn't learn it. It seems that most males (excepting homosexuals) feel the same. Unfortunately, some individuals take this repulsion to extremes, to the point of denying the whole LGBT community their civil rights, some to the point of becoming quite violent. Why is this revulsion overriding our logical minds? Bigotry certainly has a nurturing component. However, those individuals *on the other team,* people who are homosexual, are such because of their genetics. It is not a matter of choice. This trait does have a biological origin.

If there were no strong urges to reproduce, the species would end. Young boys often think that kissing girls is yucky, but, when puberty sets in, game change. Why is sexual coitus so enjoyable (to most)? When we males see females in certain situations, we fantasize and seek opportunities to further our desire. We love sex. What is it about homosexuals? Why aren't they attracted to females in the same way straight men are? Is their brain wired differently? Some people seem to think that they choose to be gay, that they are making an immoral choice that government should discourage. Seems obvious to me: we're programmed for heterosexual relations; at least 90% of us are so programmed. The other 10% received an alternate program. In 1948, Alfred Kinsey, in his book *Sexual Behavior in the Human Male,* reported that 10% of the male population is gay. The actual percentage is debatable, but it is certainly not zero. [28]

Several studies also have shown that homosexuality tends to run in families. The identical twin of a gay male is likely to be gay as well. The pedigree of pairs of gay brothers found that homosexuality tends to run on the maternal side of the family tree: the brothers had a higher than average number of maternal nephews and uncles who are gay. The same holds true for females; the sister of a lesbian is likely to be a lesbian as well. [29]

There have been numerous studies that have shown physical differences in the brains of homosexuals and heterosexuals. One of the first published scientific papers to report these differences was by the neuroscientist, Simon LeVay. In 1991, while at The Salk Institute, LeVay published a research article that received national attention. This article reported a difference in the average size of one small structure within the hypothalamus that influences gonadotropin secretion, maternal behavior, and sexual behavior in several mammalian species. This structure was more than twice as large in heterosexual men as in homosexual men. In homosexual men this structure was the same as that of women. This suggested that homosexuality has biological underpinnings. [30]

At the Stockholm Brain Institute, researchers studied the brain scans of 90 gay and straight men and women and measured the size of the two symmetrical halves of the brains. Heterosexual women have symmetrical halves of similar size. Heterosexual males have slightly larger right hemispheres. Gay males' brains resemble female brains in size, and lesbian females have brains that resemble those of heterosexual males. The number of nerves connecting the two sides of the brains of gay men were also more like the number in heterosexual women than in straight men. In other words, structurally, the brains of gay men were more like heterosexual women, and gay women more like heterosexual men. [31]

The preference for a male or female sex partner is largely determined by embryonic exposure to sex steroids. In the embryo, during the period of sexual differentiation, cells are influenced by their genetic composition and their hormonal environment. Specific genes induce gonadal differences in appropriate tissues. Other genes induce structures in the brain that are important for sexual orientation be it hetero-, homo-, bi, or transexual. Major endocrine changes during embryonic life often result in an increased incidence of homosexuality. [32, 33]

A male mouse's desire to mate with either a male or a female is determined by the brain chemical serotonin, a neurotransmitter governing sexual preference in mammals. Serotonin is known to regulate sexual behaviors, such as erection, ejaculation and orgasm, in both mice and men. Neuroscientist Yi Rao of the National Institute of Biological Sciences in Beijing and his collaborators have shown that serotonin also underlies a male's decision to woo either a female or another male. [34]

There are some people who just hate the taste of Brussels sprouts. If they were to be served Brussels sprouts on the plate they would push them aside uneaten. There are some other people that just love this vegetable, especially if sautéed with a bit of balsamic vinegar. There is a genetic reason for these two sets of traits. There are genes that determine what chemicals or foods we can actually taste and that are pleasantly favorable. With respect to Brussels sprouts, the distastefulness that some humans experience is the presence of a gene called TAS2R38. Food choice is partially related to a person's ability to taste and hence is genetic.

There are biological differences in their taste receptors between people; not everyone's taste receptors are the same. Not everyone loves Brussels sprouts.

Brussels sprouts belong in the plant genus *Brassica*, which includes cabbage and mustard greens. Plants in this genus contain a chemical called 6-propyl-2-thiouracil (PROP).

There is another closely related chemical to PROP, called Phenylthiocarbamide or PTC. For some individuals PTC tastes bitter; others do not taste the chemical at all. PTC does not occur in food but PROP and related chemicals do. The history of genetic linkage of tastes began in 1931 in the DuPont laboratory by a chemist named Arthur Fox. Fox accidentally released a cloud of a fine crystalline PTC. Standing near Fox was a colleague who complained about the bitter taste in the air. Fox, who was standing much closer to this cloud of PTC, tasted nothing. Further research revealed that about 70% of people can taste this chemical; 30% are non-tasters. Before DNA testing, PTC was used in paternity tests.

Homophobia may have a genetic component. It may not be entirely our nurturing. I do not believe there is any scientific research to substantiate this notion, but consider this reasoning. Obviously heterosexuals have a genetic trait that prompts them to seek out and mate with members of the opposite sex. The biology of one's sexual preference may include some genetic component that includes a "distaste" of having sex with one's own sex, or for some homosexuals a distaste of having sex with the opposite sex. Bi-sexuals may not have this punitive distaste gene.

Considering that the distaste of some foods and odors is genetic; homophobia may also be genetic in a fashion similar to but distinct from the genetics of distaste of certain foods and odors. Of course one cannot smell homophobia, and it is certainly not the TAS2R38 gene of Brussels sprouts. However there are likely other genes, unrelated to the taste buds on the tongue or the olfactory receptors in the nose, that predispose one to have a "distaste" of the inappropriate sex. Prejudice against homosexuals may have a genetic link.

To my knowledge there have not been any crusades banning lovers of Brussels sprouts. Non-Tasters do not march in Sprouts Pride Parades, while Tasters taunt and disparage these *queer* people. Tasters are not dragged behind pickup trucks by some non-taster bigot. All this may seem rather silly, but serious injustices have been carried out on members of the LGBT community by individuals with silly ideas in their heads. Prejudices (and silliness) can be over-come with information, tolerance, understanding and compassion.

A large percentage of the general population feels compelled to ostracize the gay community. Biblical references are often stated. People of deep faith want to act as the moral police, similar to the Islamic fundamentalistic Taliban in Afghanistan. With these people ancient writings seem to outweigh the scientific literature.

*Prejudice - a vagrant opinion without visible means of support.*

—Ambrose Bierce (1842-c.1914), an American editorialist, journalist, short story writer, fabulist, and satirist.

There is a great deal of worldwide righteous indignation towards homosexuality. One church, the *Westboro Baptist Church* is very active protesting gay associated events. There have been Christian ministries that have the compulsion to set these gays straight, a practice called *Gay Conversion.* Why does a religious group seek out gays for conversion to straightness? Is it a mission, a personal crusade? Perhaps they envision that converting gays is a path to their own redemption and the glories of heaven. There is a great deal of sin in the world, but straightening out the gay community attracts their attention. You have probably noticed an occasional newspaper article that exposes preachers caught with pornography or with prostitutes, both male and female. Then it is revealed that they have spent an inordinate amount of time preaching against these sins. This is very strange. Do you suppose that *Gay Straighteners,* people who attempt to straighten gays may themselves have sex-orientation issues?

A Christian group that once promoted therapy to encourage gays and lesbians to overcome their sexual preferences has closed its doors and apologized to homosexuals, acknowledging its mission had been hurtful and ignorant. *Exodus International*, the oldest and largest Christian ministry dealing with faith and homosexuality, had been operating since 1976. On June 19, 2013 it announced on its website that they would cease operations. The

Irvine, California-based group's board unanimously voted to close Exodus International and begin a separate ministry. President Alan Chambers issued a public statement that said, "I am sorry for the pain and hurt that many of you have experienced. I am sorry some of you spent years working through the shame and guilt when your attractions didn't change. I am sorry we promoted sexual orientation change efforts and reparative theories about sexual orientation that stigmatized parents. Chambers said he was part of a System of Ignorance." [35]

How can it be that there are homosexuals in the human species as well as in many others, when they have no chance of propagating? The genes of non-propagating individuals are lost. There seems to be no positive selection for such a trait.

Charles Darwin, in his 1859 book The Origin of Species, reflected on the puzzle of sterile social insect, such as honey bees. Males are sterile and it's the females that reproduce. A selection benefit to related organisms would allow the evolution of a trait that confers the benefit but destroys an individual at the same time. A trait that favors the reproductive success of an organism's relatives, even at a cost of the organism's own survival and reproduction, would be an evolutionary strategy. John Maynard Smith is credited for coining the actual term *kin selection* in 1964. E.O. Wilson in his 1975 book *On Human Nature* suggests that it is kin selection that can explain the continued presence of homosexuality.

As an alternative, I offer this explanation. In humans the probability that each fertilized human egg survives to be born is rather rare. Studies by John Opitz, Professor of Pediatrics, Human Genetics, and Obstetrics and Gynecology at the University of Utah, have shown that between 60% to 80% of all naturally conceived embryos are simply flushed out in women's normal menstrual flows unnoticed. The rate of natural loss for embryos that have developed for seven days is 60%. The total rate of natural loss of human embryos increases to at least 80% if one counts from the moment of conception. [36]

Conformational research has found that 66% of all human embryos fail to develop successfully. It is estimated that 50% to 67% of all human conceptions do not develop successfully to term. Many of these embryos express their abnormality so early that they fail to implant in the uterus. Others implant but later fail to establish a successful pregnancy. Most abnormal embryos are spontaneously aborted, often before the woman even knows she is pregnant. [37]

In a developing embryo there are an immense number of biochemical reactions that must occur in the proper sequence for success, including the embryo's hormonal environment. Non-viable embryos simply lack the biological ability to develop into anything resembling a viable baby. Things can and often do go astray. With respect to homosexual individuals, abnormal biochemical events may misdirect the embryo's sexual orientation to be in opposition to the overall phenotype of the individual. As the frequency of homosexuality is relatively frequent, say 5% to 10%, viability of those embryos with this misdirected sexual orientation must survive equally well; orientation is not a factor for embryo survival. If homosexuality can be attributed to atypical biochemical events of the sexual orientation pathway and there is no positive selection for heterosexuality over homosexuality, one could conclude that the frequency of homosexuality will remain relatively constant within the population. This assumes that these atypical biochemical events continue to occur at a constant but low frequency.

Another example of events that can go wrong during embryonic development is the birth of children with indeterminate sexuality, known as intersex individuals. These children have a mixture of male and female chromosomes and genitalia. The frequency of such an event is one child per 2000. Very often surgery is performed to correct the outward appearance. In one case, a person with no clear gender-defining genitalia was subjected to surgery. The person said many years later: "I am neither a man nor a woman. I will remain the patchwork created by doctors, bruised and scarred." Germany has become Europe's first country to allow babies with characteristics of both sexes to be registered as neither male nor female. Parents are now allowed to leave the gender blank on birth certificates, thus creating a new category of *indeterminate* sex. [38]

Incest is forbidden in most all societies, with a few exceptions. We know that Egyptian royalty practiced incest as method of maintaining family lines. In most societies however, incest is an appalling act. Are there genetic traits that repeal one against incest, or is it all in our nurturing? While no anti-incest gene has been described, there would be no selective advantage for such unions, as the offspring often suffer severe genetic defects. Repulsion of such behavior is a strong emotion. All emotions originates in the brain. What, if any, genetic components might be involved is not known, but is reasonable to assume that there are genetic components in addition to any nurturing that affect such behaviors.

There are also taboos of sexual relations with pre-pubescence children. Is this cultural or are there some genetic factors involved? Both incest and pedophilia do occur in societies, but the majority of society abstain from these behaviors. Practitioners of these behaviors seem to understand that it is illicit but feel an uncontrollable compulsion. Knowing the negative consequences, why do they do it? I would assume that there are genetic components involved in these behaviors, for both abhorrence and for obsession or temptation. Would these taboos have any genetic similarities to the abhorrence of homosexual relations that most heterosexuals manifest? Of course, genetic factors do not excuse such behavior, but they might offer explanations.

Humans are basically alike the world over. We are of one species. Just as skin color is a genetic trait, so is one's sexuality. It is likely that we are programed for disdain of sexual encounters with individuals opposite to our genetics. Humans cannot choose how they, as embryos, develop, nor can they change their genetics at some later stage in life. Physical appearance, religiosity, ethnicity, political or sexual orientation are factors that are the focus of people that hold prejudices. [39] Most certainly nurturing is the major factor; we learn prejudice as well as tolerance from our parents and acquaintances. But humans can make conscious choices as to how to regard others.

The genetic underpinnings for the trait of prejudices have not been established. If there is genetic linkage, most likely it would not be one specific gene, but a multiple of genetic factors responsible.

## Trait: *Memory - False Memory*

*When I was younger, I could remember anything,*
*whether it had happened or not.*

—MARK TWAIN (1835-1910)

Humans are certainly selective in their perception of the world. We have selective hearing, selective sight and selective memory. We know that children have selective hearing. Remember when your mother called out to you to wash up and come to dinner, just when you were in the middle of your favorite game? You continued playing. She called again this time she just said, *The pizza is ready.* Within four seconds you were at the table; however you didn't stop to wash up. Our perceptions are certainly influenced by our environmental experiences.

*What you see is not what is really there; it is what your brain believes is there. Seeing is an active, constructive process. Your brain makes the best interpretation it can according to previous experience and the limited and ambiguous information provided by your eyes.*

—-FRANCIS CRICK [40]

Our vision is selective. We drive along the road only seeing what is important for safe driving. The color of the cars around us go unnoticed. We pay attention to only what we deem important. We are oblivious to the rest of the world unless there is an immediate and unexpected change in our world. We assume the we perceive the world as accurate as a camera. Unfortunately we have selective perception. Witnesses to a bank robbery give multiple different accounts as to the description of the robber. There is false memory, when we "remember" what didn't occur. The human brain is not as perfect as we may believe. All the sensory signals we receive are processed within our brain and are subject to the physical and biochemical processes of the brain along with our prior memories.

*There are lots of people who mistake their imagination for their memory.*

—Josh Billings, the pen name of the nineteenth-century American humorist Henry Wheeler Shaw (1818-1885).

Human brains are of sufficient size to store decades of memories with sufficient storage space for future memories. There is not a specific location in the brain that is the memory center. The memory process is brain-wide, a complex web of neurons spread across relevant cortices. Our senses receive signals that are encoded and transmitted to the appropriate areas of the brain: the frontal, parietal, temporal or occipital lobes, as appropriate. Critical to memory storage is the amygdala.

When we observe a particularly beautiful flower, we look at and smell the flower, feel the summer's breeze and hear birds chirping. Various appropriate areas of the brain receive encoded messages. We consolidate new related experiences with older experiences such as related encoded experiences of viewing flowers of a previous springtime outing. What seems to be a single memory is a complex construction. Recalling a memory effectively reactivates the neural patterns generated during the original encoding. Short-term and long-term memories are encoded and stored in different ways and in different parts of the brain. [41]

To study memory formation the common laboratory roundworm *Caenorhabditis elegans* is often used. Hundreds can live in a petri dish and they only require bacteria as a source of food. Salk Institute scientists using *C. elegans* found that activation of specific neurons lead to increased signaling from a protein called CREB (cAMP response element binding protein), a cellular transcription factor important to learning. CREB binds to DNA at a specific locus, resulting in an increased or decreased activation of genes essential for learning and retaining new memories. The presence and the amount of CREB protein determines how quickly an animal learns. These studies could lead to new avenues of research for brain enhancement. [42]

Memory can be manipulated. The memory of a rat can be selectively removed and predictably restored by stimulating nerves in the brain at frequencies that are known to weaken or strengthen the connections between nerve cells called synapses. By stimulating a specific neuron in the amygdala, a brain region known to be essential for fear conditioning, *We can form a memory, erase that memory and we can reactivate it, at will, by applying a stimulus that selectively strengthens or weakens synaptic connections,* said Dr. Roberto Malinow. His laboratory at UC San Diego, examines how neuronal activity controls the strength of communication between neurons at sites called synapses. He has found that a change in the number of synaptic receptors and long-term maintenance of this enhanced number of synaptic receptors is a critical mechanism underlying synaptic plasticity. Such synaptic plasticity is thought to underlie the formation and storage of memories. [43]

*There are three side effects of acid: enhanced long-term memory, decreased short-term memory, and I forget the third.*

—Timothy Leary (1920-1996), an American psychologist and writer known for advocating psychedelic drugs.

One of my most memorable professors at UCSB was my Biochemistry Professor, Dr. Henry Nikada. Very occasionally, when a fellow student asked a question, the answer to which he didn't know or remember, he would scratch his head, ponder and then say, *Damn if I know.* At the next session he would have the answer. He was self-assured enough to admit he didn't know or remember everything and wise enough do some research to find the answer. For this honesty and wisdom, he is fondly remembered.

*The difference between false memories and true ones is the same as for jewels: it is always the false ones that look the most real, the most brilliant.*

—-Salvador Dali (1904-1989), a prominent Spanish surrealist painter.

I find it interesting that some religious have absolute certainty concerning their religious convictions. Their convictions cannot be shaken, even when confronted with information that contradicts their beliefs. We know from our own experience that our memory often fails or there is false recollection. There are many factors that can occur during witnessing, encoding and retrieval of the event, any of which may adversely affect the creation and maintenance of that memory.

Scientists at UCSD and MIT have created a false memory in a mouse and have provided detailed clues as to how such memories may form in human brains. They caused mice to remember being shocked in one location in their cage when in reality the electric shock was delivered in a completely different location. Physical traces of a specific memory can be identified in a group of brain cells as it forms and activated later by stimulating the same cells. They identified and chemically labeled cells in specific areas of the animals' brains responsible for memory. Dr. Tonegawa, a lead scientist in this study, said that because the mechanisms of memory formation are almost certainly similar in mice and humans, part of the importance of the research is "to make people realize even more than before how unreliable human memory is," particularly in criminal cases when so much is at stake. [44]

Dr. Elizabeth F. Loftus, a cognitive psychologist at the University of California, Irvine, is an expert on human memory. Dr. Loftus is best known for her groundbreaking work on the misinformation effect, eyewitness memory and the creation and nature of false memories including recovered memories of childhood sexual abuse. She has conducted extensive research on the malleability of human memory. Memory is susceptible to misinformation, and our ability to recall events, our eye-witness account, may be flawed, concluding that our memory is volatile. In a laboratory study, Dr. Loftus has found that 25% of human subjects came to develop a memory for the event that had never taken place. Dr. Loftus has testified and advised courts about the nature of eyewitness memory for many cases, including the trials of the McMartin preschool, O.J. Simpson, Ted Bundy, Oliver North and the officers accused in the Rodney King beating.

Incidentally, in November 18, 2010, *The Bronowski Art&Science Forum* hosted, as co-presenters, Professor Loftus and Los Angeles artist Deborah Aschheim in a forum entitled *Lest We Forget - Memory.*

Why is *The Truth* so difficult to establish, especially relating to human recall? We send people to jail, even to death row based upon eye-witness accounts. Occasionally the sentence is overturned based upon DNA evidence. It has long been speculated that mistaken eyewitness identification plays a major role in the wrongful convictions of innocent individuals. Mistaken eyewitness identification is responsible for more convictions of the innocent than all other factors combined. *The Innocence Project* is a national litigation and public policy organization dedicated to exonerating wrongfully convicted people through DNA testing and reforming the criminal justice system to prevent future injustice. They have been responsible for 329 exonerations. 75% of the 329 DNA exoneration cases had occurred due to inaccurate eyewitness testimony. It appears that our memory may be tainted by the need for resolution, justice or revenge.

## Trait: *Compassion, Empathy and Love*

> *The human failing I would most like to correct is aggression. It may have had survival advantage in caveman days, to get more food, territory or a partner with whom to reproduce, but now it threatens to destroy us all. We need to replace aggression with empathy, which brings us together in a peaceful loving state.*
>
> —Stephen Hawking (b. 1942), British theoretical physicist, University of Cambridge. He is the author of A Brief History of Time.

Empathy and compassion, in common usage, are similar. Empathy is the ability to recognize, understand and share emotions that are being experienced by another. *I feel your pain.* Compassion is sympathetic pity and concern for the sufferings or misfortunes of others. Humanitarians are compassionate.

The practice of compassion is admired, but limited. We pass by a beggar, a vet obviously suffering from mental problems, a hitchhiker, all people in need. A few of us stop to assist, others stay on their journey. We make decisions as to whom we extend compassion. It seems obvious that we can't help everyone, we can't hand out a few dollars to all that need it. If we did we would be just as broke. Sometimes we are a bit afraid that the homeless person or that the hitchhiker will rob us. Sometimes we wonder why they don't just get their act together and find a job. We are reluctant to give them some money. We rationalize that they will just spend it on booze or drugs. Sometimes we're heartless.

There is a compassionate human activity: *Practice random acts of kindness and senseless acts of beauty.* There is a sense of warmth when practiced; it is sometimes infectious. This phrase is attributed to Anne Herbert (b. 1952) an American writer. She is also known for the phrase, *Libraries will get you through times of no money better than money will get you through times of no libraries.* There is another form of compassion, of unknown origin: *Pay it forward.* It is the expression for describing the beneficiary of a good deed repaying it to others instead of to the original benefactor.

> *"I do not pretend to give such a deed; I only lend it to you.*
> *When you … meet with another honest Man in similar*
> *Distress, you must pay me by lending this Sum to him;*
> *enjoining him to discharge the Debt by a like operation, when*
> *he shall be able, and shall meet with another opportunity.*
> *I hope it may thus go thro' many hands, before it meets*
> *with a Knave that will stop its Progress. This is a trick*
> *of mine for doing a deal of good with a little money.*
>
> —Benjamin Franklin in a letter to
> Benjamin Webb, April 25, 1784

There are genetic factors that seem to determine one's degree of compassion and empathy. Researchers of Emory University recorded brain activity, using fMRI, on research participants while they played a game called *Prisoner's Dilemma*. The game is designed to visualize which parts of the brain are involved when participants are given choices involving altruistic decisions, when rewards are not immediately apparent. Participants were given the chance to trust and cooperate with each other, or betray each other for immediate gain. Mutual cooperation was the most common outcome in games. The scans revealed activity in areas of the brain associated with reward anticipation, decision-making, empathy and pleasure, the amygdala. Reciprocal altruism activates a reward circuit and this activation may often be sufficiently reinforcing to override subsequent temptations to accept but not reciprocate altruism. Simply stated, when one has shown empathy one's amygdala displays activity and one feels rewarded. Those individuals with low amygdala activity do not receive any sense of reward. [45]

Empathy is an instinctive mirroring of other's experience. In the human brain, neural systems that are active when we are in pain become engaged when we observe the suffering of others. Recent brain imaging studies using f MRI have implicated parts of the limbic system in the empathic response to another's pain. [46]

Research has also shown that spending money on other people may have a more positive impact on one's happiness than spending money on oneself. [47]

Studies by Berkeley psychologists have shown that compassion is related to one's social status and apparently not genetic. Seven studies using experimental and naturalistic methods reveal that upper-class individuals behave more unethically than lower-class individuals. Upper-class individuals were more likely to break the law while driving relative to lower-class individuals. Upper-class individuals proved more selfish in an economic game. Individuals with greater wealth and education are worse at recognizing the emotions of others and less likely to pay attention to people they are interacting with. These researchers conclude who individuals that are motivated by greed tend to abandon moral principles in their pursuit of self-interest, reasoning that increased resources and independence from others cause people to prioritize self-interest over others' welfare. Upper-class individuals perceive greed as positive and beneficial. Unethical behavior in the service of self-interest that enhances the individual's wealth and rank may be a self-perpetuating dynamic that further exacerbates economic disparities in society. [48]

The question not addressed in these studies is whether there are any genetic components that differ between the upper-class and lower-class individuals. Upper-class individuals may be upper-class due to lower amygdala activity or they have genetic variants that would tend to increase one's greediness.

*Education leads to enlightenment. Enlightenment opens the way to empathy. Empathy foreshadows reform.*

—Derrick A. Bell (1930-2011), the first tenured African-American Professor of Law at Harvard Law School.

Oxytocin is a hormone and a neurotransmitter. Genetic variations in the oxytocin pathway influences social and emotional states. Normal oxytocin levels have been shown to relate to romantic love, parent-child bonding, empathy, and subsequent generosity towards strangers. Abnormally low levels of oxytocin are associated with stress, depression, lower empathy, self-esteem and optimism. [49]

Levels of oxytocin are controlled, in part, by their receptors. The receptors for oxytocin can differ, one individual to the next, resulting in behavioral differences. One study identified a very small difference in the DNA in the coding region of this receptor protein that alters behavior. The differences involved only two DNA base pairs, adenine (A) and guanine (G). Individuals with the GG sequence normally have higher levels of oxytocin. They also have the highest amygdala activity when processing emotional information. They are prosocial, emotionally stable and have the ability to easily discern the emotions of others. AG and AA individuals are more sensitive to stress, have fewer social skills, low self-esteem, are less optimistic and have greater mental issues than GG individuals. AA individuals had the lowest amygdala activity when processing emotional information. [50]

AA or AG in oxytocin receptor —> Low Levels of Oxytocin —> low Amygdala activity —> stress, depression, lower empathy, self-esteem and optimism.

> GG in oxytocin receptor —> Normal Levels of Oxytocin —> high Amygdala activity —> romantic love, parent-child bonding and empathy.

Empathy is tricky. If we spend an inordinate amount of time focused on one compassion- associated event, other ongoing worldly events will go unnoticed. Endless hours of single event TV coverage overshadows the hundreds of youth shootings in Chicago. Earthquakes are a daily earthly events, yet only a few are newsworthy. Victim association determines where we spend our empathy. If there is no commonality with the victim, there may be no empathy. Empathy is parochial, it narrows our outlook. When deciding the significance of any event, it may be better to make analytical calculations rather than to respond instantaneously with empathy.

*Where in the brain is love located?* asked the artist David Hockney of neuroscientists, Drs. Francis Crick and Tom Albright at the time of his visit to The Salk Institute in 1992. Crick and Albright looked at each other for a brief moment, then in unison responded: *The amygdala.*

What is Love? A most difficult question. What is known is that empathy, oxytocin and the amygdala are involved. A research study found that when individuals perform behaviors associated with compassion and love such as warm smiles, friendly hand gestures, affirmative forward leans, their bodies produce more oxytocin. Being compassionate causes a chemical reaction in the body that motivates us to be even more compassionate. [51]

Who would have thought that love is just an electrical-biochemical neuronal assemblage of mental manipulations? Poets, however, have other descriptive explanations.

*Part of the reason men seem so much less loving than women is that men's behavior is measured with a feminine ruler.*

—Francesca M. Cancian, American author and therapist.

*Love and compassion are necessities, not luxuries.*
*Without them, humanity cannot survive.*

—Dalai Lama XIV, *The Art of Happiness*

## Trait: *Hope*

*He that lives upon hope will die fasting.*

—-Benjamin Franklin (1706-1705)

Daniel A. Helminiak, a Catholic priest and theologian defines the human spirit as a function of the brain; it's an awareness, insight, understanding and judgment. It is doubtful that a specific location of the human brain could be identified as the human spirit loci. The term is useful and widely used to explain a certain sort of hope that humans possess; something that we hold as worthy and is our better nature. He refers to it as *The Higher Component of Human Nature.* [52]

*Hope in reality is the worst of all evils because*
*it prolongs the torments of man.*

—Friedrich Nietzsche (1844-1900), German philosopher, cultural critic, poet, composer, and Latin and Greek scholar. He wrote several critical texts on religion, morality, contemporary culture, philosophy, and science.

Hope is a prevalent human trait. There must be hundreds of thousands of songs and poems that deal with hope. Hope is the currency of religion. Hope is the state that promotes the desire of positive outcomes related to events and circumstances in one's life or in the world at large. Hope and related emotions are a field of study in Psychology. Dr. Barbara L. Fredrickson, a Professor at the University of North Carolina, argues that religious leaders are peddlers of hope. "Hope comes into play when our circumstances are dire; when things are not going well or at least there's considerable uncertainty about how things will turn out … hope literally opens us up and removes the blinders of fear and despair and allows us to see the big picture thus allowing us to become creative and have belief in a better future."[53]

Can hope "remove the blinders of fear and despair and allows us to see the big picture" ? Science examines the Big Picture; it explains the how of the universe, not the why. Fear and despair arise often when we cannot comprehend events. Hope is a placebo. Science offers an understanding based upon biological and physical principles. We can accept the findings of science, or ignore science and look elsewhere for explanations. Religion is the usual alternative. Science does not offer hope, only explanations that withstand close scrutiny and are verifiable.

*Science is a very human form of knowledge.*

—Jacob Bronowski (1908-1974), British mathematician, biologist, historian of science, theatre author, poet and inventor. He was a resident Fellow at The Salk Institute from 1964 to 1974. He is best remembered as the presenter and writer of the 1973 BBC television series and the book, *The Ascent of Man.*[54]

## Trait: *Morality*

*The hottest places in hell are reserved for those who, in times of great moral crisis, maintain their neutrality.*

—Dante Alighieri, c. 1265–1321, *Inferno*

Morality is a genetic trait that is passed through generations by way of our DNA. Surprised? Researchers at the *Infant Cognition Center* at Yale University, Paul Bloom, Karen Wynn, Kiley Hamlin and other colleagues have published in peer-reviewed journals some very interesting studies. In their laboratory, commonly called *The Baby Lab*, scientists have observed that babies are born with a sense of morality. In unique studies working with pre-verbal babies, some as young as 3 months and using puppets, the researchers have been able to quantify the preferences of these babies. Babies too young to verbalize can fix their eyes on their preference and babies a bit older, can reach out for their preference. Given the choice of good puppet and bad puppet acting out scenarios, babies chose the good puppet overwhelmingly. This finding indicates that babies are drawn to the nice guy and repelled by the mean guy.

In another experiment of 6 to 10 month-old infants, measuring helping versus hindering situations and using inanimate objects, a yellow square helping a blue circle up an incline and a red triangle pushing the circle down, infants prefer a helpful character to a neutral one and prefer a neutral character to one who hinders.

Starting at about 1 year of age, babies also seem to want to lessen the pain of others. They soothe others in distress by stroking and touching or by handing over a bottle or toy. In an experiment that may suggest an inborn genetic basis for racial bias or prejudice, 3-month-olds prefer the faces of the race that is most familiar to them to those of other races. For those who are offended by the word race in describing human characteristics, I use the word in a cultural sense, not as a scientifically meaningful term. It seems that babies possess certain moral foundations, the capacity and willingness to judge the actions of others, some sense of justice, and the inclination to respond to altruism and selfishness. [55]

Children as young as 18 months of age are able to show altruistic behavior, helping others. Cooperative behavior has also been shown to exist in chimpanzees, but to a lesser degree. This would add support to cooperation as having a genetic component. [56]

## Trait: *Imagination*

*Logic will get you from A to B. Imagination will take you everywhere.*

—Albert Einstein (1879-1955) [57]

When I was about 7 or 8 years of age, living in Monterey, Saturday afternoons were often spent at the movie theater. My friends and I would gather to see Western movies, black and white at the time. After the movie was over we would saunter out of the theater and onto Alvarado Street, each of us emulating the ambling or strutting of the cowboys we had just seen. We held our hands at our sides, pretending to have our six shooters at the ready. We certainly had imagination. We were able to produce fanciful images in our brains embracing the heroic antics of our movie heroes.

Our perceived world is an interpretation of data arriving from the senses; as such, it is perceived as real by contrast to most thoughts and imaginings. Imagination is the conscious ability to form new images and sensations in the brain that are not perceived by our five senses. It is central to many creative and uniquely human abilities. Imagination helps make knowledge applicable in solving problems and is fundamental to integrating experience and the learning process. It is central to many creative and uniquely human abilities, including the intellectual advancement in the humanities and the sciences.

One might suppose that imagination is a trait that makes *Homo sapiens* human, that which sets us apart from other animals. When this trait in our evolutionary history occurred is unknown, as well as which animals possess this trait. It is doubtful that chimpanzees have the same imaginative qualities as modern man. The development surely increased as our brains increased in size. Imagination and self-awareness were probably an early form of cognition, the ability to plan and foresee the future. *Where will the game animals be tomorrow? If I take a shortcut over to the watering hole, I'm sure they will show up a little later.* The ability to anticipate future events, to be able to consider most all consequences, have been most important for the survival of *Homo sapiens.*

Imagination is the faculty or action of forming new ideas, images or concepts of external objects beyond our senses. The imaginational process has been studied by scientists at Dartmouth College. Using multivariate pattern analysis of fMRI on 16 human subjects, the scientists measured brain activity by detecting associated changes in blood flow. In the study, test participants were asked to imagine precise visual figures, to mentally disassemble them or mentally blend them while their brains were observed for specific areas of activity. The scientists were able to track the sequence of neural activity that occurred as the manipulative tasks were performed. Their findings reveal a widespread cortical and subcortical network that operates on visual representations in the mental workspace. Researchers revealed that a widespread neural network performs specific mental manipulations on the contents of visual imagery. While revealing important glimpses, they conclude that they did not know how the human brain mediates complex and creative behaviors such as artistic, scientific, and mathematical thought. [58]

## Trait: *Curiosity and Creativity*

*The artist's job is to be a witness to his time in history.*

—Robert Rauschenberg (1925–2008), American painter, sculptor and graphic artist, well known for his *Combines*, in which non-traditional materials and objects were employed in innovative combinations.

We are a curious species. Our trait of curiosity, our desire to know or learn something, is a major factor in our ability to learn and retain information and can facilitate creativity. A recent study, published in the journal Neuron has shown that when we are curious, the brain's chemistry changes. There is an increase in activity in the hippocampus, which is involved in the creation of memories. This area of the brain is also associated with rewards we obtain, money, sex or fond foods. When the circuit is activated, our brains release dopamine which gives us a high. Dopamine plays a role in enhancing the connections between cells that are involved in learning. When we are curious, we are actually motivated, which leads to more effective learning experiences. If some fact is, to us, boring we are less likely to remember it. The converse is also true; interest in a fact equates to better memory retention. [59, 60]

Creativity arises out of curiosity. Creativity can be characterized as the process of producing something that is original, worthwhile, expressive and imaginative. Human's creativity is expressed in literature, poetry, philosophy, music, science and the arts, even religion. Art is but one aspect of humans' creativity. It an endeavor that is more familiar to me than other human expressions of creativity and as such will have greater emphasis in this book.

A tree falling in a forest does not make a sound unless there is an organism present capable of detecting vibrating air molecules and performing a translation.

Art critic and philosopher Arthur C. Danto ( 1924-2013) examined art history from a discerning perspective. Danto discussed Duchamp's conscious effort to question aesthetic values, good or bad taste. Exemplary is Duchamp's readymades, one of which is the famous white porcelain urinal displayed as such and titled Fountain. Art historians consider Duchamp's Fountain a major landmark in 20th-century art. Danto theorizes that works of art are *Embodied Meanings. Because of works like Warhol's Brillo Box, I could not claim that aesthetics is part of the definition of art.* [61]

Art can create lasting memories that share commonalities with other individuals. With just the mention of two photographs taken decades ago, it is likely that these will elicit in your brain a detailed recollection of the events well beyond the description I shall provide. The first is the photograph by of a lone student standing down four tanks in Tiananmen Square, Beijing, China, 1989 taken by Jeff Widener. The other photograph taken in 1972 by Nick Ut, is of a 9 year-old Vietnamese girl, Kim Phue, running naked after napalm was dropped on her village. These images are likely stored in the long-term memories of millions of people. We may not know the names of the photographers nor the names of the individuals, yet these iconic images resonate. I would argue that these photographs are works of art. These photographs have embodied meanings.

Over 300 Prehistoric Cave Paintings sites have been discovered in Europe, India, North and South America, Africa, Australia and Southeast Asia. The first recognized early artwork was discovered in the Blombos Cave, an archaeological site located about 300 km east of Cape Town, South Africa. The cave contains Middle Stone Age deposits currently dated at between 70,000 and 100,000 years ago. Discoveries in this cave include thousands of pieces of ochre with grid or crosshatch patterns, dated to some 70,000 years ago. Some of these recovered ochre pieces have been deliberately engraved or incised and it is suggested that they represent a kind of early abstract or symbolic depiction and are arguably among the most complex and clearly formed of objects claimed to be early abstract representations.

This suggested to some researchers that these early *Homo sapiens* were capable of abstraction and the production of abstract or symbolic art. Also found there were marine gastropod shells that were deliberately pierced through the aperture, probably with a bone tool. The evidence suggests that these shells were strung, perhaps on cord or sinew and worn as a personal ornament.

The oldest known European cave paintings were discovered in Cantabria, northern Spain, in the El Castillo cave and have been dated to be more than 40,000 years old. This cave is well known for its long archaeological and artistic sequence. Current studies in El Castillo have estimated that there are nearly 2,500 images, signs, or human-made marks. [62]

There is debate who the artists were, *Homo sapiens* or *Homo heidelbergensis,* and what their purpose was. These caves showed no habitation and the paintings were located in areas deep within the cave. There is speculation that they may have had some sort of communicative, religious or ceremonial significance. Other instances of which we are familiar are the cave paintings of Lascaux, France, which date to about 15,000 BCE.

*The aim of art is to represent not the outward*
*appearance of things, but their inward significance*

—Aristotle (384–322 BCE)

At some point in the history of man, the concept of ownership must have arisen. *This is mine.* My personal and unsubstantiated view is that art grew out of the necessity of early hominins, possibly *Homo habilis,* or *Homo erectus,* a million or more years ago to claim ownership of one's tools, perhaps by making a mark on the handle of an axe or the shaft of a spear. Maybe just a X hash mark on his handle or shaft to distinguish his tool from another's to claim ownership and authorship. Another individual might have made two hash marks to distinguish his tool from any other. A third individual possibly made a triangle on the handle. Each could then identify one's tool. The idea of claiming ownership may have had its beginnings with a simple mark. What's left after the triangle? More triangles, or squares leading to more detailed patterns. That primitive tools and weapons were "owned" by individuals is a conjecture. One could imagine that tools were manufactured to the specifications of a specific individual, for example the length of a spear appropriate for the height of that individual. Engraved symbols could indicate ownership. Symbols have embedded meaning.

Was this the origin of art, something beyond an accidental or random pattern? Was the purpose strictly a method of coding one's own possessions? *It's mine and I can prove it.* The act of marking an object with the intention of declaring ownership would not have occurred until the trait of self realization, selfishness and ownership arose as a human trait. Very early art may have been marks made in the dirt or sand that might have indicated directions, or perhaps marks were made upon rocks or boulders that have long disappeared due to weathering. All these marks would have embedded meanings; they would be art if we accept Danto's definition.

When marks became representational is when we might consider the beginning of humans making art. It would also affirm that art ultimately triggers some kind of transcendence that can only be completed by the viewer; taking a spoken word and translating symbolically into a mark on a physical object. Not any mark could be considered art without the conscious mind knowing its purpose. An accidental mark on a stick or wall or any object is not art. Intentionality, meaning and provenance need be associated with the mark. Works of art are Embodied Meanings. If we accept this interpretation as to what art is, I would suggest that the origin of art and the origin of writing occurred simultaneously, that there is a blur between early man's markings and written language. Early marks would have meanings. They could be simply called early written communication, or early visual expressive creativity.

The origin of recognized art occurred at least a half million years ago. Recently discovered were enigmatic zigzag markings on an ancient shell made between 430,000 and 540,000 years ago. These Java shells with deliberate enigmatic zigzag geometric engravings were made by *Homo erectus*. [63]

Jackson Pollock (1912-1956), an American abstract expressionist, produced what were labelled Drip or Action Paintings. Working in an old barn on Long Island, New York, he had a unique and radical approach to painting. Pollock purchased large yardage of yachting canvas from his local hardware store and spread it across the floor of his barn. Using a paint brush, not in the traditional fashion, he dipped the brush into commercial house paint and dribbled the paint off the brush directly onto the canvas. The brush never touched the canvas. At times he would take the can of paint and pour the paint onto the canvas. The uniquely continuous paint trajectories served as "fingerprints" of his motions through the air.

Research has shown that the poured patterns of Jackson Pollock are fractal, a collection of repeating patterns, that has the same statistical character as the whole. To address the long-term appeal of Pollock's work, scientists conducted an investigation of human response to fractals, as depicted in Pollack's works. The investigations include eye tracking, visual preference, skin conductance, and EEG measurement techniques. Research indicated that perception is determined by the edge contours of the observed fractal patterns in free-viewing situations, that is, subjects fixate on definite contours. In one experiment they showed that stress levels, as measured by skin conductance, were reduced, an indication of a relaxed state when an individual was observing fractal patterns such as in Pollock's paintings. His paintings induced the observer to engage their eyes in a constant search for fractals, a perceptual, physiological and neurological exercise involving cross-over brain activity. Further, they suggested that there might be a peak shift effect involved in observing Pollock's works. [64]

One of Pollock's friends, Ruebin Kadish, noted, "I think that one of the most important things about Pollock's work is that it isn't so much what you're looking at, but it's what is happening to you as you're looking at his particular work." [65]

There are numerous cognitive approaches attempting to describe the thought mechanisms of creative thinking. Alice Flaherty, professor of neurology and psychiatry at Harvard Medical School, has proposed that the creative drive results from an interaction of the frontal lobes, responsible for idea generation, the temporal lobes for idea editing and evaluation and high dopamine levels from the limbic system, which increases general arousal and goal-directed behaviors and reduces latent inhibition. [60] In other words and simplified for non-neuroscientists, our creative drive is a complicated electro-chemical labyrinthine of brain activity.

Dr. V. S. Ramachandran, Director of the Center for Brain and Cognition at the University of California, San Diego, has written a book about the neural basis of art. This book provides a framework for understanding aspects of visual art, aesthetics and design. "Our knowledge of human vision and of the brain is now sophisticated enough that we can speculate intelligently on the neural basis of art and maybe begin to construct a scientific theory of artists experience." [66]

Ramachandran, together with his collaborator William Hirstein, discuss *Peak Shift,* a well-known principle in animal discrimination. Peak shift is the principle that there is increased interest in a desired object when the object increases its size. In the peak shift effect, animals sometimes respond more strongly to exaggerated versions of the training stimuli. If a rat is taught to discriminate a square from a rectangle of say 3:2 aspect ratio and is rewarded for the rectangle, it will soon learn to respond more frequently to the rectangle. Paradoxically, however, the rat's response to a rectangle that is even longer and skinnier, say of aspect ratio 4:1, is even greater than it was to the original prototype on which it was trained. The rat's brain makes the leap to interpreting the appearance of greater rectangularity as being several times better. Taking this phenomenon of hyper-normal stimuli further toward art and why we respond as we do, Ramachandran and Hirstein wonder, *Is our brain hardwired to appreciate art?*

When you look at any evocative object, image or picture, the image is extracted by the early visual areas and sent to the inferotemporal cortex, an area specialized for detecting faces and other objects. Once the object has been recognized, its emotional significance is gauged by the amygdala at the pole of the temporal lobe, and if it is important, the message is relayed to the autonomic nervous system by way of the hypothalamus. Your response may be a sense of joy or flee or mate, depending upon prior experiences and circumstances. "Some artists deliberately exaggerate creative components such as shading, highlights, and illumination to an extent that would never occur in a real image in order to produce a caricature. These artists may be unconsciously producing heightened activity in the specific areas of the brain in a manner that is not obvious to the conscious mind....Tapping into the figural primitives of our perceptual grammar and creating ultra-normal stimuli that more powerfully excite certain visual neurons in our brains as opposed to realistic-looking images." [67]

Martin Skov, a Danish neuro-aesthetician and Oshin Vartanian, Professor of Psychology, University of Toronto, investigated the neurobiological underpinnings of aesthetic preferences using fMRI to determine areas of enhanced functional connectivity during active art practice. The areas determined to be active when viewing paintings was distributed between the parietal and frontal cortices. "These results suggest that viewing paintings triggered activity in regions of the brain associated with visual representation, object recognition, pleasure in addition to systems that underlie the conscious processing of new information to give it meaning." [68]

*An artist is not paid for his labor but for his vision.*

—James Whistler (1834-1903), American-born, British-based artist. His art was characterized by a subtle delicacy, while his public persona was combative.

Lawrence Weiner (b. 1942), a conceptual artist, was interviewed by NY Times journalist Roberta Smith. "Driven by the joy of language and quite a bit of humor, Mr. Weiner's ebullient work asks tough questions about who makes or owns art, where it can occur and how long it lasts, It reminds us that while art and money may have been inextricably entwined throughout most of history, art's real value is not measured in strings of zeros, high-priced materials or bravura skill, but in communication, experience, economy of means (the true beauty) and, yes the inspired disturbance of all status quo. It also affirms that art ultimately triggers some kind of transcendence that can only be completed by the viewer." Mr. Weiner has elevated Robert Rauschenberg's famous dictum to the effect that "This is art if I say so, to the more inclusive this is art if you think so. His polymorphous efforts create situations in which such thoughts feel not only natural, they feel like our own." [69]

There may be no consensus as to what constitutes art and what falls outside the circle. It does not matter all that much. It does seem that many humans feel a compulsion to express themselves often by writing, painting, even doing craft-work. When one is in the act of art practice, outside distractions seem to vanish. It usually requires deep concentration. For some reason, a deep feeling of accomplishment and appreciation is associated with completion. This desire to create symbolic images is not restricted to the visual arts. Poetry, music, even architecture are humanities, they are creative arts. In the presence of an art creation, some sort of transcendence occurs that is completed by the witness.

*If you are only moved by color relationships, you are missing the point. I am interested in expressing the big emotions; tragedy, ecstasy, doom.*

—Mark Rothko (1903–1970), American painter of Russian Jewish descent. He is generally identified as an Abstract Expressionist. With Jackson Pollack and Willem de Kooning, he is one of the most famous postwar American artists. His works have been referenced as having a sublime spirituality, creating a state of silence and reflection.

## Trait: *Tribalism*

*No tribe unites with another of its own free will.*

—Sir Arthur Keith (1866 - 1955), Scottish anatomist and anthropologist.

Tribalism is characterized as having a tendency to form groups, with strong group loyalty. A supportive group with commonalities such as culture, social or economic status, religion or kinship can be considered a tribe.

The term tribe refers to both ancient groups of humans as well as present-day groups with similar commonalities. The derivation of the word *tribe* comes from the Latin *tribus*, referring to the three ethnic divisions of Rome, established by Romulus c. 750 BCE.

Early hominins certainly existed within tribes. Tribal existence provided the hub for learning and passing on tracking and hunting skills, knowledge of the prey's habits and techniques to kill the prey. The skills of the gatherers, identifying edible plants and their the habitat, were tribal essentials. It is likely the elderly assisted in raising and teaching the young. The tribe was where everyone slept and fed together; safety in numbers. This close proximity of individuals may have favored selection of individuals with enhanced forms of the trait of empathy, identifying with and comforting those with pain.

Tribal benefits included greater opportunities for mating, grooming, heat conservation and food. A species that is instinctively kind to their own tribe members and could judge the moral character of others has an evolutionary advantage. The altruistic nature of tribal animals is comprehensible when considering the benefits of sharing goods and services. Non-sharing individuals stand alone. Our basic morality may have been one factor, a factor that seems to be genetically determined, passed on through our genes from one generation to the next. Without compassion, empathy and cooperation we would not have been a successful tribal animal. There are costs to individuals within a tribe. The chance of receiving parasites or other pathogens increases with the number of members, as does intra-tribal conflict and reproductive competition.

Robin Dunbar (b. 1947) is a British anthropologist and evolutionary psychologist and a specialist in primate behavior at the University of Oxford. Dunbar has calculated optimal tribal size of early humans referred to as *Dunbar's number.* The bases of this calculation is the relationship of primate brain size with their tribal size and then extrapolating to early *Homo sapiens.* He proposed that humans can only comfortably maintain 150 stable relationships. Tribes larger than this generally require larger long-term memory, more restrictive rules, laws, and enforced norms to maintain a stable cohesive group. As tribes increase in size, splintering would likely occur. When a splinter tribal group is isolated with no interbreeding with the original tribe, some mutations may occur that may be advantageous in a new environment. Splintering and occupying different territories with different environments and different challenges may have been the causative agent in the evolution of various hominin species.

In contemporary times the term tribal remains meaningful. It is used as a term for groups with cultural or other similarities. Different tribes can be distinguished by skin color, religion, language, politics or other commonalities. Tribalism is very likely a biological trait. The array of genes that contribute to our tribal nature are not fully understood.

There are two major human tribes that are familiar to us with each having different mindsets, principles or personal constitutions. They are classified as conservative and liberal. These terms refer to behaviors, moralities, attitudes and practices. These terms are often but not necessarily associated with religious and political groups.

As an adjective, being conservative often means being cautious and holding to traditional attitudes. Conservatives can be sober and conventional, avoiding novelty or showiness. They are disposed to preserving existing conditions and institutions.

Liberals are open to new behavior or opinions and are willing to discard traditional values. They are favorable to progress or reform. Liberals tend to be freer from prejudice or bigotry.

Fear is one trait that binds tribes. Fear of the consequences if the other tribe has greater power, that the other side will cause havoc.

There are hundreds of tribal associations: Republicans, Democrats, Libertarians, Catholic, Baptists, Musslims, Jews, Rastafarians or other affiliations. We may also be members of the Sierra Club, the PTA, the NRA, the Rotary Club and any other numerous clubs and affiliations. Most likely we belong to several tribes. Tribal behavior often involves flying the colors, using bumper stickers or dressing to reflect one's tribal alliance. For some it might be designer outfits, for others, outdoor apparel. Tattooing has become quite common and for some it may be tribal symbolism We support in a very tribal fashion our favorite sports team. For some it may be wearing a wedge-shaped cheese hat. We echo our tribe's sentiments: *No new taxes, No Muslims allowed, Protect the Polar Bears and the Rainforest, Give Peace a Chance.* The last phrase is lovely but we rarely practice peace. We seem to always be at war.

We seek friendship from like individuals and shun outsiders. This is not a trait learned from our parents. It's natural. We are tribal. We may not form herds, except when there is a Super Bowl or similar event. We can be extremely loyal tribe members. There is empirical evidence that it is in our nature to form stable interpersonal relationships. People form social attachments readily under most conditions and resist the dissolution of existing bonds. The need to belong is a powerful, fundamental and extremely pervasive motivation. Belonging is a basic survival strategy. [70]

There are genetic traits that would be critical for tribal unity. The genetic traits of empathy, compassion and cooperation would be essential. Oxytocin would certainly be central for tribal formation. The oxytocin system arose very early in animal evolutionary history. Certain peptides found in early invertebrates, such as gastropods, snails and slugs, share commonality with human oxytocin, vasopressin and related peptides. These gastropods may not have empathy but they do have peptides that are similar to the peptides that constitute human neuropeptides. [71]

Other studies have demonstrated linkage of oxytocin to behaviors of ethnocentric behavior, trust and empathy of in-groups and suspicion and rejection of outsiders. [72]

Charles Darwin, in his book The Descent of Man, a discourse on human evolution, covers societal issues concerning evolutionary theory, ethics, psychology, sexual selection, as well as differences between human races. Darwin postulated that *Homo sapiens* succeeded because of their traits of compassion and sharing, traits that would benefit the whole tribe. Wealth-sharing assured that the tribe survived.

Cooperation is instinctual and was a major factor in the survival of *Homo sapiens*, says Michael Tomasello, an American psychologist and

Co-Director of the Max Planck Institute for Evolutionary Anthropology in Leipzig, Germany. He has synthesized three decades of published research to develop a comprehensive *Evolutionary Theory of Human Cooperation.* Children can be cooperative, apes not so much. If you were to drop something in front of a two-year-old the child would be likely to pick it up and hand it to you. Through observations, Tomasello argues that children are instinctively cooperative. Cooperation is not necessary for the apes. When early humans moved out onto the grasslands, they scavenged the carcasses of game animals that had previously been killed. This was the primary source of their protein. Cooperative behavior as distinct from the fierce aggression between chimp groups, was the turning point that shaped human evolution. [73]

## Trait: *Consciousness*

The smell, the color and the texture of a rose are all processed in different parts of the cortex. Yet our experience is bound together into one cohesive, conscious experience. The mechanism of consciousness is poorly understood. It remains mystical to some degree. Two brilliant scientists, Francis Crick of The Salk Institute, together with his long-time collaborator Christof Koch of the California Institute of Technology, postulated that the *claustrum* was the key to consciousness. The claustrum is a thin sheet of grey matter that lies concealed beneath part of the cortex, the outer covering of the brain that carries out the computations involved in seeing, hearing and language. Most, if not all, regions of the cortex have two-way connections to the claustrum, as do the structures involved in emotion. A wealth of information converges in the claustrum and leaves it to re-enter the cortex. The claustrum may be involved in widespread coordination of the cerebral cortex, using synchronization to achieve a seamless timescale both between the two cortical hemispheres and between cortical regions within the same hemisphere, resulting in the seamless quality of conscious experience. Drs. Crick and Koch postulated that the claustrum is similar to a symphony conductor, bringing all the sounds of the various instruments into one harmonious sound. The different parts of the cortex must bring the inputs of sight, smell, touch, etc, into one experience. We are conscious in one combined experience. [74,75]

Psychedelic drugs are known to produce hallucinations and altered perceptions of reality. Australian scientists have analyzed thousand of reports of individuals who had consumed *Salvia divinorum* and, for comparison, LSD. *Salvia divinorum*, commonly called Diviner's Sage, is native to Mexico. The Mazatec civilization's priests would chew its leaves to get in touch with the gods. *S. divinorum* acts on the kappa-opiate receptors. These receptors are structurally similar to the mu-opiate receptors of LSD, which mainly bind serotonin neuromodulator receptor proteins. The kappa-opiate receptors are concentrated in the claustrum and, at lower concentrations, the frontal cortex and the amygdala. The activity of *S. divinorum* likely inhibits the claustrum via its activation of the kappa-opiate receptors. Consuming the *Salvia* plant might just cause the inactivation of the claustrum necessary to test Crick's and Koch's hypothesis. [76]

I will return to the question: *Do we have Free Will?* When tempers flare, when we choose certain stocks to purchase, when we become sexually aroused, when we are inspired by a work of art, are we operating freely or are there genetic factors involved? Our behavior is a complex interaction of our ancient genetic traits as well as our acquired knowledge and experiences. It is certain that we don't have complete Free Will, nor are we automatons. We certainly are tribal animals on this planet.

# Five

## Behavior

*Human behavior emerges from the interaction and interplay of genes, socialization and environmental stimuli, working through ontogenetic neurobiological processes embedded in an evolutionary framework.*

—Theodosius Dobzhansky (1900-1975), prominent Ukrainian-born American geneticist and evolutionary biologist. [1]

Throughout history and even today, there have been a few humans who have been practitioners of horrendous, atrocious and immoral acts. And there are other humans who live their lives with high ethics, beyond reproach. We are the only species capable of attaining high levels of atrocities. We have the intellectual capacity to invent and implement schemes that create pain and suffering. We have the mental skills to recognize grotesque pain in others. Many believe that human behaviors are a consequence of our nurturing - how we were raised, our experiences and the circumstances in which we find ourselves. These factors do have essential roles in human behavior. Less considered are humans genetic traits. Anytime we act, the brain with all its complexity is involved. Any behavior is extremely complex and to assign just one trait would be naive. As yet, it cannot be determined precisely which traits may be involved in humans complex behaviors. What follows are illustrations of man's ignoble as well as man's noble behaviors.

## Behavior: *Slavery*

Ingenious was the fellow who, millions of years ago, discovered how to make rope. Rope created opportunities for early humans. The first sort of rope was probably made of vegetation, vines, long palm leaves or sliced slivers of animal skin woven together. Rope had many uses. Rope could lash sticks together as part of the structure of a shelter. It could be the handle of baskets to ease hauling goods. By lashing together the front legs and the rear legs of a killed animal, then slinging it over a pole, tribe members could share the load back to their home.

Rope also could have enabled early hominins to bind into subservience captured members of a hostile tribe or another hominin-competing species. This discovery would have enabled the tribe to take hostages and force them to do the menial work. Slavery most likely arose early on the path to modern humans.

Slavery is arguably the most ignoble act of human kind. The history of slavery spans nearly every culture, nationality and religion from ancient times to the present day. Seymour Drescher (b. 1934), an American historian and a professor at the University of Pittsburgh, writes of slavery,: "The most crucial and frequently utilized aspect of the condition is a communally recognized right by some individuals to possess, buy, sell, discipline, transport, liberate, or otherwise dispose of the bodies and behavior of other individuals." [2]

Modern slavery may be a much larger problem than recently estimated. Nearly 36 million people worldwide or 0.5% of the world's population live as slaves, as stated in the *Global Slavery Index*. The index considers both children and adults to be "living in slavery" if they have been victims of sex or debt bondage, mandated marriages or forced labor. [3]

## Behavior: *Torture*

*The wish to hurt, the momentary intoxication with pain, is the loophole through which the pervert climbs into the minds of ordinary men.*

—Jacob Bronowski

The nature of humans would need to include the history of torture. Torture is, what most intelligent humans would consider, a very dark element of humankind. Slavery must have predated recorded history and continues to the present.

The purposes of torture seem to have several agenda: to extract confessions from prisoners, to punish or to cause fear to anyone contemplating doing a forbidden act. Often the sadistic act brings some sort of pleasure to the torturer or his leader.

Methods of torture have continually improved, so to speak. Methods have included those with which you may be familiar, one hopes through reading. Included here is a small selection of the "ingenious" methods man has concocted to torment others. Examining one's history may give a perspective that could alter future events by avoiding making the same mistake.

*The Rack,* a torture device wherein the victim's limbs are strapped to a device that allows the torturer by means of pulleys very gradually, in steps, to increase the tension on the limbs inducing excruciating pain. Eventually the victim's bones are dislocated with a loud crack, caused by snapping cartilage, ligaments or bones. If the torturer kept turning the handles the limbs would eventually be torn off.

One of the early written accounts of use of the rack is that of Epicharis (d. 65 BCE). She was a freed slave and a key figure in a conspiracy attempt to assassinate Emperor Nero, the Emperor who "fiddled while Rome burned". The plot reflected the growing discontent among the ruling class of the Roman State with Nero's increasingly despotic leadership. When Epicharis was exposed, Nero ordered her to be tortured because she refused to name any of her co-conspirators. Epicharis was beaten, burned with fire and tortured on the rack. Her tormentors could not extract any confession from her. When, on the second or third day, she was carried in a sedan chair, for her limbs were already broken, to be tortured a second time, she strangled herself on her way by using her girdle, which she had fastened to the chair. Tacitus, a senator and Roman historian said of her, that she was more noble than many others who without being tortured would betray their nearest relatives.

*The Inquisition,* which started in the 12th century in France, was a judicial system within the Roman Catholic Church set to combat heresy and religious sectarianism. The rack was extensively used as a method of combatting the folly of the unfaithful. Except within the Papal States, the institution of the Inquisition was abolished in the early nineteenth century.

*The Iron Maiden* was first used in the Ming Dynasty (1368–1644). The disloyal prisoner would be locked into an iron container that was suspended over hot coals. The executioner would pour water onto hot coals. The steam was a certain and painful death.

*Dunking* was widely used during The Tribunal of the Holy Office on the Inquisition commonly known as the *Spanish Inquisition.* The Tribunal was established in 1482 by King Ferdinand and Queen Isabella, famous for financing the 1492 voyages of Christopher Columbus. Dunking was used to convert Muslims and Jews to the Christian faith. During the Salem Witch trials, between February 1692 and May 1693, dunking was a punishment for sorcery and usually "favored" women.

King Henry VIII shared in the history of torture. In 1532 he made boiling a legal form of capital punishment.

Other creative methods of doing away with the unwanted ones was freezing, by forcing a victim to stand outside, in winter, naked, with cold water poured on their head. Another method was live burial; the victim was buried up to his neck, followed by stoning, or pouring honey on them and allowing ants or other animals to feast.

*Waterboarding* is a form of torture in which water is poured over a cloth covering the face and breathing passages of an immobilized captive, causing the individual to experience the sensation of suffocation and drowning.

In the *Age of Enlightenment,* a period of western intellectual history (the late 17th to the late 18th centuries), the practice of waterboarding was banned.

The populace found it morally repugnant. Unfortunately it didn't disappear. Darius Rejali, a professor at Reed College in Oregon and author of the book, Torture and Democracy, said waterboarding has been an interrogation technique preferred by the world's democracies, including past White House officials. It is an attractive interrogation technique, as it causes great physical and mental suffering yet leaves no marks on the body. [4]

In recent years, it wasn't merely low-level brutalizers and their immediate superiors who sanctioned and approved torture techniques but senior White House officials, including National Security Adviser Condoleezza Rice and Vice President Dick Cheney. To spruce up the term to sound very appealing, they changed the term to *Enhanced Interrogation.* In addition to waterboarding these "compassionate conservatives" condoned the use of hypothermia, stress positions, electroshock and intimidation by barking snarling dogs.

Dick Cheney has repeatedly said that waterboarding is not torture. On NBC's *Meet the Press,* December 14, 2014, Cheney said, given another chance to authorize such methods, *I'd do it again in a minute.* This is after the CIA's report that no actionable information was generated by torture. Cheney's no-regrets attitude is chilling, enhanced by the unsettling fact that so many Americans agree with him. Immediately after the CIA report, a poll by *The Washington Post-ABC News* found that six in ten Americans approved of the CIA's treatment of suspected terrorists.

The argument against waterboarding often cites Abu Zubaydah. The CIA, convinced that he was harboring knowledge of future attacks, subjected him to twenty days of torture. The FBI refused to take part. They waterboarded him 83 times, stripped him, deprived him of sleep, slammed him into the prison wall and played music at deafening volumes. He began releasing information after only 35 seconds. It was later found that he was lying and making up information to stop the torture. The CIA interrogators said that, if Zubaydah died during questioning, his body would be cremated. If he survived the ordeal, the interrogators wanted assurances that he would "remain in isolation and incommunicado for the remainder of his life." Condoleezza Rice told the CIA that these harsher interrogation tactics were acceptable. [5] In September 2009, the Obama administration acknowledged, during Abu Zubaydah's habeas corpus petition, that Abu Zubaydah had never been a member of al-Qaeda nor been involved in the attacks on the African embassies in 1998 nor the attacks on the United States on September 11, 2001. The US Government has not officially charged Zubaydah with any crimes. According to the *NY Times*: The Guantanamo Docket his current status: *Held in Indefinite Law-of-War Detention and not Recommended for Transfer.* As of Jan. 14, 2016, he has been held at Guantánamo for nine years and four months. [6] This is an example of both its uselessness as a method of obtaining reliable information and the depths that some humans will go to for *the momentary intoxication with pain.*

Is torture a genetically determined behavior of humans. Or do mothers teach their children the fine art of torture? Is Dick Cheney deficient in serotonin? Does he have high levels of dopamine or low levels of GABA? Perhaps it's heightened levels of fear, anger, aggression, delusion, irrationality, greed and prejudice or just his irrational obsession with torture. Is he lacking in moral character and empathy? Perhaps he suffers a psychopathic disorder. To say that Dick Cheney is an animal just might impinge upon the character of all the other animals in the world. Unfortunately, there is a plethora of Dick Cheney-esque humans. Humans are capable of inventing and implementing the most horrific methods to dominate other humans.

## Behavior: *Cleansing*

*This is what men do when they aspire to the knowledge of gods.*

—Jacob Bronowski, as he stood in front of the remains of the crematoriums at Auschwitz and dips his hands in the pond containing the ashes of some of the people murdered there, who included members of his own family.

The most atrocious behavior in the history of mankind must be Nazi Germany's extermination policies, "The Final Solution". Six extermination camps were established in former Polish territory, where six million ordinary

European Jewish people were enslaved as part of their war effort, often starved, tortured and killed.

Far too common are policies of governments throughout history that have resorted to campaigns to segregate and "cleanse" their tribes of what they consider undesirables. The US government set up internment camps for the Cherokee and other Native Americans in the 1830s. During World War II, 110,000 Japanese/Americans were "interned" in camps across the USA. The reason: their appearance was physically similar to the Japanese who attacked Pearl Harbor. The conservative mindset at the time was that of fear without the use of logic, reason or compassion.

Apartheid was a system of racial segregation in South Africa from 1948 to 1994. Nelson Mandela (1918-2013) is credited as being the inspirational leader who led the ending of apartheid. In 1962 he was arrested, convicted of conspiracy to overthrow the government and sentenced to life imprisonment. He served 27 years in prison. In 1986, in reaction to apartheid, the United States Congress passed and sent to President Ronald Reagan the *Comprehensive Anti-Apartheid Act,* which banned bank loans and the trade of numerous goods. Ronald Reagan vetoed the bill, but it was overridden by the Republican-controlled Senate.

The motivations for ethnic cleansing have been stated as: "they" have a different god and it is our *God's Will* that we do away with them, or "they" speak a different dialect or "they" control all the banking or they stole our lands 14 centuries ago. There are numerous justifications that induce the wrath of some humans. Often any excuse will be used. Fear, aggression and revenge are in full blossom.

Some perpetrators of torture and cleansing may be mere underlings carrying out the orders of higher-ups. They may have no choice but to obey or face similar consequences; these people, in their ordinary home environments, are quite peace-loving individuals.

*The play of tolerance opposes the principle of monstrous certainty that is endemic to fascism and, sadly, not just fascism but all the various faces of fundamentalism. When we think we have certainty, when we aspire to the knowledge of the gods, then Auschwitz can happen and can repeat itself. Arguably, it has repeated itself in the genocidal certainties of past decades.*

—Simon Critchley (born 1960), English philosopher, commenting on the words of Jacob Bronowski.

*It is said that no one truly knows a nation until one has been inside its jails. A nation should not be judged by how it treats its highest citizens, but its lowest ones.*

—Nelson Mandela

Some practices of imprisonment are a not too subtle form of cleansing. There certainly is a need to incarcerate some individuals. Many of the incarcerated began life as moral individuals but, due to circumstances, poor upbringing, broken homes, lack of education and the unavailability of decent jobs, they end up in the prison system. Worldwide there are a great variety of prisons, from the more lenient with goals of rehabilitation to the severe, used as a facility to inflict punishment.

The United States is number one in the world in pro rata incarceration. With less than 5% of the world's population, the United States has nearly 25% of the world's prisoners. There are 2.24 million people in prison in the US, about triple the number locked up in the 1980s. Nearly 1% of Americans are in prison, the majority due to drug offenses, mainly marijuana. One in every three black males born today can expect to go to prison at some point in his life compared, with one in every six Latino males and one in every 17 white males. Racial minorities are more likely than white Americans to be arrested. Once arrested they are more likely to be convicted, and once convicted they are more likely to face stiff sentences. There are 3,281 prisoners in America serving life sentences for nonviolent crimes with no chance of parole. A prison sentence, for most, is the lifelong loss of voting rights. [7]

For-profit corporations such as *Corrections Corp of America* and *GEO Group* are the leaders in the correction industry. It's a $70 billion industry.

These corporations spend millions of dollars lobbying States and Congress to stiffen sentencing. Privatization of our prison system has the intended purpose of lowering costs. The "unintended" consequence of all penal systems is suppressing the black and Latino vote. A form of political shenanigans, "modern" slavery or bigotry perhaps? [8]

Mass incarceration creates one of the major public health challenges facing the United States. Millions of people cycle through the courts, jails and prisons every year. In our prison system there are far higher rates of chronic health problems, substance use and mental illness than the general population. [9]

With 2.24 million incarcerated people in the US, hardships fall on the families they leave behind. "The extravagant brutality of the American approach to prisons is not working. If you treat people badly, it's a reflection on yourself." [10]

Contrary to the United States, Sweden will be closing four prisons since there has been a drop in their prison population. Two factors have been attributed to the decline. One is Sweden's liberal prison approach, with its strong focus on rehabilitating prisoners. The other factor is that Swedish courts have given more lenient sentences for drug offenses. Is there a lesson to be learned? [11]

In Norway, there is no death penalty, there are no life sentences. The maximum term for any crime is 21 years. The goal of the Norwegian penal system is to get inmates out of the system. The recidivism for Norway is 20%. In the US 68% of released prisoners are re-arrested for a new crime within 3 years. 76% are re-arrested within 5 years. [12]

*The degree of civilization in a society can*
*be judged by entering its prisons.*

—Fyodor Dostoyevsky (1821-1881), Russian novelist, short story writer, essayist, journalist and philosopher. Dostoyevsky's literary works explore human psychology in the troubled political, social, and spiritual atmosphere of 19th-century Russia.

## Behavior: *War*

*War does not determine who is right, only who is left.*

—Bertrand Russell (1872-1970), British mathematician, philosopher, prominent anti-war and anti-imperialism activist and winner of the 1950 Nobel Prize for literature. He went to prison for his pacifism during World War I. He considered himself a liberal, a socialist, a pacifist and an atheist. His attitude was that war results from an impulse of aggression. *War grows out of ordinary human nature. The only remedy is to find an equal and opposite passion, such as love.* [13]

Are we not a peaceful, loving and compassionate species? *Give peace a change.* Peace has never had a chance. Why? Because we are *Homo sapiens* and warring is our nature. We have seen bumper stickers that state "Pray for World Peace" a world where there are no wars. This seems to be everyone's wish. What if humans never ever waged war, no one was sent off to war, no one was killed in war? Consider only a few of the USA's wars: Civil War, 620,000 soldiers died; World War I, 16,000,000, died; War War II, 60,000,000 people died. In just these three wars, more than 76,000,000 individuals died.

Not included are the Korean War, Vietnam, Gulf "wars" or other police actions. Not included are the countless wars over at least 10,000 plus years, thousands of wars. The number of young men and women killed in all wars must be in the high hundreds of millions, maybe billions. These were men and women who likely did not breed. Their contribution to the world population had they not died but lived and bred would have added billions more people to our planet. The world population would be at least 10 times greater that today's population of 7.2 billion inhabitants. War: a blessing for population control advocates.

So, in a demonic sense, war is good for the human species; it keeps or at least slows humans from over-populating the earth. Along with population control, war increases employment. The Military Industrial complex, remember? Without hundreds of overseas military bases, without Halliburton and similar corporations along with the Armed Services, unemployment would soar. The wars of humans have been a huge economic success. I guess the take-home lesson is: War is good. So what will your response be when next you see that bumper sticker, *Give peace a change*?

We humans may be warriors. We may inflict pain needlessly. We may be excessively selfish. Are the congressional Militaristic Interventionists genetically predisposed to wage war? Are these individuals just posturing? Is their hawkishness similar to a peacock display, an eagerness to copulate or an eagerness to invade? I don't have the answer; just a thought.

In the Sunday, February 13, 2011 installment of Gary Trudeau's comic strip *Doonesbury*, Mark Slackmeyer, a longtime character in the strip who is a liberal radio host makes a point about gun violence.

*"What are we like as a people?"* Slackmeyer muses to himself in his studio. *"Nine years ago we were attacked -- 3,000 people died. In response, we started two long, bloody wars and built a vast homeland-security apparatus -- all at a cost of trillions! Now consider this. During those same nine years, 270,000 Americans were killed by gunfire at home. Our response? We weakened our gun laws."*

## Behavior: *Destruction - Altering the Climate*

The biosphere is the global sum of all regions of the surface, atmosphere and hydrosphere of earth. It includes all ecosystems; all life forms. It is a closed system, apart from solar and cosmic radiation and heat from earth's interior. It had been self-regulating for over 4 billion years. *Homo sapiens* are now the new regulators. With a nod to Walt Kelly's Pogo, *We have met the enemy and he is us.*

In 1750, at the beginning of the Industrial Revolution, the concentration of carbon dioxide in the atmosphere was 280 parts per million. In 1960 the level was 320 ppm; the 2006 level equaled 381 ppm. *The National Oceanic and Atmospheric Administration* revealed on May 6, 2015, that, for the first time in recorded history, global levels of carbon dioxide in the atmosphere averaged over 400 parts per million (ppm) for an entire month of March 2015. That's the most in 57 years of direct measurements, a near 70% rise on pre-industrial levels and probably the highest since the Pliocene Epoch, 2.6-5.3 million years ago. [14]

Humans are adding more than 35 billion metric tons of carbon dioxide to the atmosphere each year. That is equivalent to 100 million tons of carbon dioxide per day. Converting tons to pounds, that's 200 trillion pounds of carbon dioxide per day, every day. [15] Yikes!

Our atmosphere is huge but adding 100 million tons of carbon dioxide today and another 100 million tons tomorrow, can one doubt that there would not be terrible consequences? One of the most remarkable aspects of the paleoclimate record is that there is a direct positive correlation between atmosphere carbon dioxide concentration and global temperatures calibrated over the last four hundred thousand years. Over that period there have been four peaks and valleys of carbon dioxide levels, from 200 ppm to a high of 290 ppm, which corresponds with global temperatures. The present fifth peak of 400 ppm is the highest for at least the past 10 to 15 million years. [16]

Here is a brief explanation of Greenhouse gases:

The earth's atmosphere is composed of 78% nitrogen, 21% oxygen with the remaining 1% composed of about 10 other gases. Many gases exhibit greenhouse properties. In order of abundance, the greenhouse gases in Earth's atmosphere are water vapor, carbon dioxide, methane, nitrous oxide, ozone and chlorofluorocarbons. Greenhouse gases that are of most significance with respect to climate change are carbon dioxide, which comprises about 0.04% of the atmosphere, and methane, at a level of 0.0002%. Carbon dioxide has a long lifespan in the atmosphere, taking many decades and even centuries to leave. Nitrogen and oxygen together comprise 99% of our atmosphere; however, they are transparent to infrared radiation, hence are not greenhouse gases.

The high-energy short-wavelength energy radiation from the sun in the form of ultraviolet (UV) rays and visible light passes through the mostly transparent atmospheric gases and is absorbed by the cooler earth's surfaces. This high-energy short-wavelength electromagnetic radiation heats the earth's surface, clouds and the oceans. The heated surfaces re-radiate this energy away from the heated objects in the form of longer infrared wavelength radiation back towards space. While the dominant gases of the atmosphere (nitrogen and oxygen) are transparent to infrared, the so-called greenhouse gases, primarily water vapor ($H_2O$), $CO_2$, and methane ($CH_4$), absorb some of the infrared radiation. They collect this heat energy and hold it in the atmosphere, delaying its passage back to outer space. The average temperature on earth is 14 degrees Celsius (57.2 degrees Fahrenheit). Without these greenhouse gases the earth's temperature would be about minus 19 °C (-2°F). There would be no liquid water. Life as we know it would not be possible.

The term "greenhouse gas" is familiar and convenient; however the term is a bit of a misnomer. A major part of the efficiency of the heating of an actual garden-style greenhouse is the trapping of the air so that the energy is not lost by convection, keeping the hot air from being blown or convected away. This is one benefit of a backyard greenhouse.

Carbon dioxide in the atmosphere to a large extent is the result of human activity. The carbon cycle is the balance of carbon dioxide in the atmosphere, the oceans and the biosphere, a flux of production and uptake. Natural production of this gas results from volcanic activity, the combustion of organic matter, including wildfires, and the respiration of living aerobic organisms. The leading cause of anthropogenic sources of carbon dioxide is from the burning of fossil fuels. Other sources are the industrial production of cement and deforestation. Our oceans and inland waters are a natural sink for carbon dioxide, resulting in their acidification, essentially our waters are becoming carbonated and more acidic. This acidification of our oceans is a significant problem for marine calcifying organisms like coral and some plankton becoming vulnerable to dissolution as the oceans become more acidic. As the coral reefs disappear so will most smaller fish. The whole oceanic food chain is threatened. Ocean acidification is associated with some of the worst crises in biotic history. The Permian extinction took place roughly two hundred and fifty million years ago and killed off more than 90% of all organisms on the planet. This drastic kill-off was caused by increased volcanism, which resulted in drastic increases on carbon dioxide and methane gases, ocean acidification, sea level changes and shifting in ocean circulation driven by climate change.

Methane, the second most prevalent greenhouse gas emitted by human activities, accounts for 9% of total greenhouse gas emissions. It is nevertheless a huge part of the problem. Methane traps up to 72 times more heat that carbon dioxide. In addition to its direct radiative impact, methane has a large indirect radiative effect because it contributes to ozone formation. [17]

Methane in our atmosphere predominately has a biogenic origin, a byproduct of the decay of vegetation. Cows, humans and other animals also expel methane gas as a consequence of their diet. Far more significant is that as the polar permafrost warms, methane will be released into the atmosphere as a consequence of thawed and decaying tundra. However, the far greater source of methane lies under our oceans as methane hydrate. There are roughly ten billion tons of rather stable crystals of methane hydrate on the cold ocean floor. The total amount of carbon that humans have released into the atmosphere since the start of the Industrial Revolution, 250 years ago, is 337 billion tons. As the oceans warm, these methane hydrate crystals will melt, releasing gaseous methane into our atmosphere. The methane hydrate's contribution to the atmosphere will accelerate the earth's warming. [18]

A meta-analysis of climate-induced global extinction rates has produced an alarming conclusion. A meta-analysis is a statistical technique for combining the findings from numerous independent studies. Results indicate that extinction risks will accelerate, with future global temperatures threatening up to *one in six species* under current policies. Extinction risks were highest in South America, Australia, and New Zealand and risks did not vary by taxonomic group. Without drastic changes to our climate policies, the possibility of massive extinctions is quite real. [19]

A comprehensive analysis by 90 experts of more than 35,000 surveys conducted at nearly 100 Caribbean locations since 1970 shows that the region's corals have declined by more than 50%. [20]

Australia's Great Barrier Reef, the world's largest coral reef ecosystem, has been reduced by 50% over the past 27 years. The reef is vanishing due to climate change, predatory starfish and intense cyclones linked to a warming of the oceans, according to scientists from the Australian Institute of Marine Sciences and the University of Wollongong. [21]

The International Energy Agency predicts that without efforts to stabilize atmospheric concentrations of greenhouse gas, average global temperature rise is projected to be at least 6°C within two centuries. The US Navy predicts that summers in the Arctic may be ice-free by 2016. [22]

As the Arctic ice melts, what remains is dark, less reflective ice-free oceanic water that will absorb heat, increasing arctic water temperatures. In Antarctica, as its ice melts, left exposed are rock formations which will absorb the sun's energy. As temperatures increase, water molecules are more likely to go into the vapor phase; there will be more water vapor in the air. Warm air holds a greater amount of water vapor that cooler air. Warming of the atmosphere and the oceans increases the occurrence of severe weather such as hurricanes. The oceans are warming more rapidly than at any time during the past 10,000 years. [23]

What is alarming and going mostly unnoticed is that Antarctic temperatures rose 3°C (5.4°F) in the last half century, much faster than Earth's average, said Ricardo Jana, a glaciologist for the Chilean Antarctic Institute. [24]

In late 2014, two research papers reported that the great ice sheet of West Antarctica has become irreversibly destabilized due to warming waters. The studies show that the rate of ice loss from West Antarctica is increasing, with the acceleration particularly pronounced in the past decade. This is due to warmer ocean waters pushing up from below and bathing the base of the ice sheet. [25]

In early 2015, there was a scientific paper reporting that a gigantic glacier on East Antarctica was suffering a similar fate. The Totten Glacier, which holds back a vast catchment of ice, has cavities that are causing the glacier's retreat. The researchers used three separate types of measurements taken during their aerial over-flights, gravitational measurements, radar and laser altimetry to get a glimpse of what might be happening beneath the massive glacier whose ice shelves are more than 1,600 feet thick in places. The result was the discovery of two undersea troughs or valleys beneath the ice shelf, regions where the seafloor slopes downward, allowing a greater depth of water beneath the floating ice. These cavities or subsea valleys, the researchers suggest, may explain the glacier's retreat. They could allow warmer deep waters to get underneath the ice shelf, accelerating its melting. [26]

In a research paper published in March, 2016, noted climatologist James Hansen together with 18 other international climatologist have concluded that if global temperatures on our planet continue to go up, ferocious superstorms could become more frequent and sea levels could rise several meters over the next century, drowning coastal cities along the way. [27]

The Gulf Stream is a powerful, warm and swift Atlantic ocean current that originates at the tip of Florida, follows the eastern coastline of the United States and Newfoundland before crossing the Atlantic Ocean to the west coast of northern Europe. This Gulf Stream's warm water increases summer and winter temperature along our eastern seaboard and northern European countries.

What should be very worrisome to northern Americans, Canadians and Europeans is the fresh water resulting from the warming of the Arctic. The Arctic ice cap and Greenland hold enormous amounts of fresh water. As a consequence of global warming there will be a collision of the warm salty oceanic waters moving up from the south; with vast amounts of fresh, cold and denser water from the warming Arctic and Greenland ice. There will be a blockage of the Gulf Stream that will leave Europe with colder winters and summers. With a cooler growing season, food production will suffer. [28]

A 2015 scientific study predicts that as the oceans continue to accumulate heat, increasing the melting of marine-terminating ice shelves of Antarctica and Greenland, a point will be reached at which it will be impossible to avoid large-scale ice sheet disintegration, with sea level rise of at least several meters. The economic and social cost of losing functionality of all coastal cities is practically incalculable. This is the conclusion of a new scientific study that considered ice melt, superstorms and evidence from paleoclimate data and climate modeling. [29]

*Climate change is perhaps the major challenge of our generation* said Michael H. Freilich, Director of Earth Sciences at NASA, one of the agencies that track global temperatures. 2014 was the hottest on earth since record-keeping began in 1880, scientists reported. Global warming has not ceased; it continues on a runaway projection. The year 2015 has been the hottest in recorded history. The year 2016 is predicted to be even warmer. [30]

*We may be sitting on a precipice of a major extinction event.*

—DOUGLAS J. MCCAULEY, ECOLOGIST, UC SANTA BARBARA. HE HAS CONCLUDED THAT HUMANS ARE ON THE VERGE OF CAUSING UNPRECEDENTED DAMAGE TO THE OCEANS AND THE ANIMALS LIVING IN THEM. [31]

In 2014, *The World Bank* released a scientific analysis of global warming, *a* 275 page report with citations of over 900 peer reviewed scientific articles. This World Bank report is an assessment of the consequences to human habitation in the absence of strong climate-mitigating policies. The report states that it is extremely likely that global temperatures will reach 4.0° to 5.2°C above pre-industrial levels within this century. However, many of the worst projected climate impacts in this report could still be avoided by holding warming below 2°C. [32]

Billions of people now live in or near coastal regions of the world. They certainly will need to move to higher ground. There are over a half billion people who live in low-lying littoral areas surrounding the Bay of Bengal, eastern India, Bangladesh, Sri Lanka, Myanmar, Thailand, Malaysia and Sumatra. [33]

Bangladesh, one of these low-lying countries has a population of 160 million and is in grave danger of flooding from rising sea levels. When the flooding occurs and great areas are below sea level, I feel confident that India will extend a warm invitation, welcoming the people of Bangladesh to come on over to India with a hearty: *Welcome! We have plenty.* Oops, sorry, the whole eastern and western edges of India will be under oceanic water as well. No doubt there will be food shortages. Will food fights lead to a nuclear holocaust, or to *just* conventional warfare? Not to be overlooked are the Island Nations, such as the Marshall Islands and the Maldives; they will need to be relocated.

Severe climatic conditions are expected to continue and worsen. As global temperatures increase, temperate forests will dry further, which facilitates insects growth that kills the trees, increasing the probability and severity of forest fires. The frequency of grasslands burning will increase, as Australia is presently experiencing. California and the West have been experiencing severe droughts. Leaders are now considering: water for farmers and ranchers or water for citizens. The Tigris and Euphrates, the Nile and most other major rivers pass through numerous countries. What action will down-river countries take when up-river countries deplete the supply?

In Africa, within the next decade, nearly 250 million humans will suffer increased water stress, famine, loss of agricultural production and sanitation problems. In Latin America there will be a replacement of tropical forests by savannah, resulting in a loss of biodiversity. Agriculture there will also suffer. Over 80% of world agriculture today remains dependent on the rains, just as it did 10,000 years ago.

The UN's Nobel-winning *Intergovernmental Panel On Climate Change (*IPCC) predicts climate change will cut food production by 2% worldwide each decade through the rest of the century. Meanwhile, global food demand is projected to rise 14% over the same period. Greater famine is to be expected. Worldwide, expect flood damage to the infrastructure, as storms increase in severity.

Overall there will be more droughts, heatwaves, floods, severe storms, fires, crop failures and famine. Contagious diseases are more likely to spread without adequate water for sanitation. Uncontrolled migration, hardships and conflicts will be likely. This crisis will be a job killer. Wars should be expected as countries with lowered agricultural production will most likely "Look" to neighboring countries for "assistance".

*India's development imperatives cannot be sacrificed at the altar of potential climate changes many years in the future. The West will have to recognize we have the needs of the poor.*

—Piyush Goyal, India's Power Minister. [34]

There is a positive aspect of global warming for investors. McKenzie Funk has suggestions as to how to profit from the Climate Crisis. Among his many suggestions is to purchase future farmland in northern climes such as within the Arctic Circle and Greenland. Be sure to secure the mineral rights. Other money-making strategies include designing floating homes and cities and buying elevated lands in Bangladesh and coastal India. Among other suggestions are financing genetic research to develop salt-tolerant crops such as rice and maize. Invest in renewable energy technologies, solar and wind. So as to not lose money, he warns don't invest in insurance companies that provide property insurance. [35]

President Obama, by using his executive authority under the 1970 Clean Air Act, proposed regulations to the Environmental Protection Agency to cut carbon pollution from the nation's power plants 30% from 2005 levels by 2030. His action has induced China, the largest current producer, into making its first serious commitments. [36]

Zheng Guogang, China's chief climatologist, made a rare official acknowledgement: Climate change could have a huge impact on the country's crop yields and infrastructure. [37]

In December, 2015 representatives of 195 nations reached a landmark climate accord in Paris. For the first time nearly every country committed to lowering planet-warming greenhouse gas emissions to help stave off the most drastic effects of climate change. The United Nations secretary general, Ban Ki-moon, said in an interview "This is truly a historic moment. For the first time, we have a truly universal agreement on climate change, one of the most crucial problems on earth." President Obama said: "This agreement sends a powerful signal that the world is fully committed to a low-carbon future. We've shown that the world has both the will and the ability to take on this challenge." This climate accord will not, on its own, solve global warming. At best, it will cut global greenhouse gas emissions by about half the level needed to stave off an increase in atmospheric temperatures of 2°C or 3.6°F The 2°C increase is the atmospheric temperature that is the tipping point where scientific studies conclude that the world will be locked into a future of devastating consequences, including rising sea levels, severe droughts and flooding, widespread food and water shortages and more destructive storms. [38]

Climatologists, tens of thousands of them, use multiple resources to take measurements for evidence of earth's warming. Scientists trek out onto glaciers and the tundra and go on oceanic explorations. There are aerial observations. They all are measuring and observing the changes that are occurring as the temperature slowly increases. They are especially interested in measuring temperatures, biomass redistribution and the chemical composition of the earth, waters and the air, all to find evidence of change. Satellites are used to measure and map ice and glacial flows, rainforest retreats and the rate of desertification. They measure radiances in various wavelength bands and then calculate temperatures of the atmosphere, land and water and map the oscillations of oceanic waters. Ice core samples from glaciers and the polar caps can reveal ancient temperatures and atmospheric compositions. Climatologists examine ancient pollen samples and limestone deposits and measure tree rings; they scour the earth looking for clues concerning the warming of the earth. They publish their research data and analyses in peer- reviewed journals. A peer-reviewed scientific report is one that has been evaluated by other qualified academics competent in that specialized field, experts that can evaluate the veracity of the research. From listening to the skeptics and deniers, one would think climatologists merely stick a moist finger in the air to check the direction of the wind and then huddle together to dream up their conclusions (conspiracy). Climatologists work very hard and are dedicated to obtaining verifiable results, not concocting conspiracies.

In a 2013 survey of over 4,000 papers there is a 97% consensus amongst the scientific experts and scientific research that humans are causing global warming. [39] If thousands of Fire studies produced by highly qualified scientists and peer-reviewed conclude that the probability of your million-dollar home burning down to the ground within the next few years was 97.2%, would you not buy home fire insurance? You could save a few hundred dollars by declining to purchase fire insurance with the assertion that these studies may be wrong or that these experts are perpetuating a hoax or we need greater consensus.

## There Are Contrarians

*We are all born ignorant, but one must*
*work hard to remain stupid.*

—Benjamin Franklin

*Two thousand years of published human histories say that*
*warm periods were good for people. It was the harsh, unstable*
*Dark Ages and Little Ice Age that brought bigger storms,*
*untimely frost, widespread famine and plagues of disease.*

—Senior Fellow Dennis Avery, Hudson Institute, American conservative nonprofit think tank based in Washington, DC.

Dennis Avery (b.1936) is the director of the Center for Global Food Issues at the Hudson Institute. He is an outspoken supporter of biotechnology, pesticides, irradiation, industrial farming and free trade. He does not believe that DDT causes egg-shell thinning in eagles. The Hudson Institute's financial backers include major agricultural companies like ConAgra and Cargill, as well as pesticide manufacturers Monsanto Company, DuPont, Sandoz and Ciba-Geigy.

158 elected representatives in the 113th Congress, all Republicans, have publicly declared that climate change is not caused by man. These same Republicans have taken over $58.8 million from the fossil fuel industry that's driving the carbon emissions which cause climate change. [40]

*I have offered compelling evidence that catastrophic global warming is a hoax. That conclusion is supported by painstaking work of the nation's top climate scientists.*

—Senator Jim Inhofe, Republican, Oklahoma.

Senator Jim Inhofe is the ranking member of the US Senate Environment and Public Works Committee. In November 2002, Inhofe compared the US Environmental Protection Agency to a Gestapo bureaucracy and the then EPA Administrator, Carol Browner, to Tokyo Rose. [41]

In December 2009 in Copenhagen, the United Nations hosted the *Climate Change Conference,* a global gathering of thousands of scientists and reporters. In attendance at this conference was writer David Corn, Washington Bureau Chief for *Mother Jones*. He interviewed Senator Jim Inhofe. In this interview he asked the Senator: "There are thousands of intelligent and well-meaning people in this gigantic conference center: scientists, heads of state, government officials, policy experts. They believe that climate change is a serious and pressing threat and that something must be done soon. Do you believe that they have all been fooled, victims of a well-orchestrated hoax?" "Yes", said Inhofe. "That's some hoax", Corn countered. "But who has engineered such a scam?" "Hollywood liberals and extreme environmentalists." Inhofe replied. "Really?" Corn asked. "Why would they conspire to scare all these smart people into believing a catastrophe was underway, when all was well?" Inhofe replied: "To advance their radical environmental agenda." "Who in Hollywood is doing this?" asked Corn. "The whole liberal crowd," Inhofe said. "But who?" "Barbra Streisand", Inhofe responded. [42] This might not be believable, except that it was witnessed and reported by the world press.

More Inhofe: As widely reported on February 26, 2015, after record snows blanketed the US Northeast, Senator Jim Inhofe, who wrote the book, *The Greatest Hoax: How the Global Warming Conspiracy Threatens Your Future,* brought a snowball to the Senate floor to prove climate change is a hoax. Inhofe took to the floor to decry the "hysteria of global warming." The Senator, well known for voicing criticism of scientific climate change literature, tossed the snowball onto the floor to make his case that global warming is a hoax. Unfortunately Inhofe is not scientifically literate. He should have known that warmer temperature in the atmosphere holds more moisture. Warm air equals high humidity and cold air equals low humidity. He confused weather with climate. He frequently cites Genesis 8:22. *"As long as the earth remains there will be springtime and harvest, cold and heat, winter and summer"* as evidence that climate change is not actually happening. [43]

The clinically recognized *Delusion Disorder* is where the patient suffers from persistent delusions that could not possibly be true and the symptoms persist for at least one month. The delusion disorder is not associated with the effects of a drug, medication, medical condition or schizophrenia. A person with delusional disorder may be high-functioning in daily life, and this disorder bears no relation to one's IQ. [44]

Top contributors to Jim Inhofe's campaigns, 1989-2014, were Oil and Gas, $443,200, Electric Utilities, $141,550, Health Lobbies, $127,775 and Koch Industries, $101,150. [45]

The Cato Institute, founded by Charles Koch, created a list of 100 climate change deniers. The Cato Institute's website states that these scientists maintain that the case for alarm regarding climate change is grossly overstated. "Surface temperature changes over the past century have been episodic and modest as there has been no net global warming for over a decade now. There has been no increase of severe weather-related events." From their website: The Cato Institute is a public policy research organization: a think tank, dedicated to the principles of individual liberty, limited government, free markets and peace.

ExxonMobil and other fossil fuel corporations have combined with traditionally conservative corporate groups such as the US Chamber of Commerce and conservative foundations like the Koch brothers' Americans for Prosperity to raise doubts about the basic validity of what is, essentially, a settled scientific truth. That message gets amplified by conservative think tanks such as the Cato Institute and the American Enterprise Institute and then picked up by conservative media outlets like *Fox News* as well as conservative websites. [46]

The Koch brothers revealed that they have contributed $79 million dollars from 1997 to 2011 to climate-denial front groups. [47]

Matt Ridley is a Conservative member of the House of Lords. He is also a British journalist, an author and a business man. On September 5, 2014, his article *Whatever Happened to Global Warming?* was published in *The Wall Street Journal.* This half-page editorial objected to climatologists conclusion that the earth is warming. Citing selectively from the *UN International Panel on Climate Change,* which stated that the earth's surface temperature rise has slowed over the past decade, he infers, incorrectly, that globally, temperatures are not rising and the flux is well within measuring errors. He fails to mention that hundreds of scientific studies have shown more global warming in the past 15 years than the prior 15 years. About 90% of overall warming goes into heating the oceans, and the oceans have been warming dramatically. Dr. Jeffrey Sachs of Columbia University's *Earth Institute* has sharply criticized Ridley for citing selected data and then drawing his own conclusions based on the limited selected data. Dr. Sachs, basing his criticism on the actual data within the Report, went on to state that the Report's conclusions are the very opposite of Ridley's. Incidentally, Matt Ridley is a proponent of fracking. Matt Ridley has been reprimanded by the House of Lords for failing to disclose appropriately his personal financial arrangement with the Weir Group, a British company that has been described as the world's largest provider of special equipment used in the process of fracking. [48]

Editorials in *The Wall Street Journal* are well known to be conservative. Unfortunately this and many other editorials in the WSJ were apparently not checked for veracity nor were opposing views published. This is disconcerting, especially in the case of global warming, considering its importance for humanity's future. One would hope that the WSJ would be *Fair and Balanced.* Perhaps it is, in the same fashion that FoxNews is *Fair and Balanced.*

There are numerous websites that present views counter to established climatological science. One of these sites, *Watts Up With That?*, claims to be "*The world's most viewed site on global warming and climate change.*" The site contains hundreds of articles that attempt to discredit peer reviewed articles published in recognized scientific journals. This site also attempts to discredit the climatologists. Articles concerning alleged conspiracies are common. None of the articles published on this website are original scientific research papers nor are they peer- reviewed. It is certainly not a website that recognized climatologists view seriously. This site does not stand alone. There are blogs and self-serving websites with the same agenda, discrediting peer-reviewed science.

Conservative groups have spent up to $1billion a year on the effort to deny science and to oppose action on climate change. The climate change counter-movement has been largely underwritten by conservative billionaires often working through secretive funding networks. An extensive study into the financial networks that support groups denying the science behind climate change and opposing political action has found a vast, secretive web of think tanks and industry associations bankrolled by conservative billionaires. [49]

*Today, the global warming alarmists are*
*the equivalent of the flat-Earthers.*

—TED CRUZ [50]

Ozone is a highly reactive molecule that contains three oxygen molecules. The Ozone Layer is found in the atmosphere 15 to 30 kilometers above the earth's surface. It acts as a shield protecting all life from the harmful ultraviolet B radiation emitted by the sun.

In 1985 a group of scientists made an unsettling discovery, a marked decrease in stratospheric ozone over the South Pole in the Antarctic. The depletion appeared during the southern hemisphere's spring (October and November) and then filled in. Soon after the Antarctic hole was discovered, Canadian scientists discovered that the ozone layer above the Arctic is also thinning significantly.

Researchers discovered that Chlorofluorocarbons (CFCs), chemicals found mainly in spray aerosols and refrigerants, was causing the ozone layer to break down and release chlorine. When CFCs reach the upper atmosphere and are exposed to ultraviolet rays, they break down into substances that include chlorine. The chlorine reacts with the oxygen atoms in the ozone and splits apart the ozone molecule. One atom of chlorine can destroy more than a hundred thousand ozone molecules. [51]

In response to these scientific discoveries, the international community of policymakers in 1987 adopted the *Montreal Protocol on Substances that Deplete the Ozone Layer*, which was designed to control the production and consumption of CFCs. Subsequent amendments to the protocol have followed detailed discoveries about the processes that produce the ozone hole and have largely eliminated anthropogenic emissions of CFCs and halons. Without the Montreal Protocol the growth of those emissions would have been disastrous for the ozone layer. Unlike fossil fuels, there was simply not enough money in CFCs to warrant politicians denying the ozone hole.

Had the CFC industry made large financial contributions to the campaigns of Washington elected representatives, as the fossil fuel industry presently does, the ozone depletion would have, no doubt, continued. Skin cancer would have killed millions. [52]

## Behavior: *Destruction - Living Organisms*

*Destroying rain forests for economic gain is like*
*burning a Renaissance painting to cook a meal.*

—E. O. Wilson, born June 10, 1929, American biologist, theorist, naturalist and author. His biological specialty is myrmecology, the study of ants. He is a two-time winner of the Pulitzer Prize for General Non-Fiction and a New York Times bestseller for The Social Conquest of Earth, Letters to a Young Scientist and *The Meaning of Human Existence.*

A report issued by the *World Wildlife Fund* analyzed the ocean's role as an economic powerhouse and outlined the threats that are moving it toward collapse. The value of key ocean assets is conservatively estimated to be at least $24 trillion. This report's focus was on the economic value our oceans represent for this planet, as the future of humanity depends on their healthy living conditions. Human destruction caused by greed, climate change and continued neglect is resulting in the collapse of the fisheries and mangrove forests as well as the disappearance corals and seagrass. These collapses are threatening the marine economic engine that secures lives and livelihoods around the world. Human population growth and reliance on the sea makes restoring the ocean economy and its core assets a matter of global urgency. [53]

Early humans practiced slash and burn, clearing an area by cutting and burning the vegetation. They would temporarily locate their settlement there until the soil's fertility declined. They would then move on to a more fertile area and repeat the process. That is the time when humans started the systematic destruction of the planet earth.

*A tree's a tree. How many more do you need to look at?*

—California Governor Ronald Reagan,
opposing expansion of Redwood National
Park, Sacramento Bee, March 3, 1966.

Easter Island, located 2,300 miles west of Chile and some 2,500 miles southeast of the Marquesas Islands, measures 14 by 7 miles. It was first visited by Europeans on April 5, 1722, Easter Sunday, by Dutch navigator, Jacob Roggeveen. He estimated the indigenous population to be about 2000-3000 inhabitants. Archeologists have given estimates that the first inhabitants arrived about 800 AD; however, there is some suggestions that they could have arrived as early as 400 AD or as late as 1200 AD. Oral history says the first inhabitants arrived in two large canoes with about 30 men, women and children, with provisions that included bananas, taro, sugarcane, paper mulberry, chickens and rats; the rats were either stowaways or were brought intentionally as a food source. The origin of the first inhabitants is thought to be the Marquesas Islands. Easter Island at the time of the arrival of the Rapa Nui, as they were known, was a subtropical island populated with at least 21 species of trees that grew up to 50 feet, including the world's largest palm trees, *Paschalococos disperta*, which grew to 65 feet with large trunks measuring 36 inches in diameter. There were six species of native land birds. By 1722 almost all these plants and birds were extinct.

The 887 giant stone statues, called *moai,* have been found all around the island. They range from 6 to 33 feet tall and weigh up to 86 tons. They were carved to resemble humans but with over- sized heads, long ears and pursed lips on top of thighs, without a trunk.

Over a period of approximately 800 years, the population increased to an estimated 15,000. They split into several settlements or tribes. By the 1500s the collapse was very evident. The forests had disappeared, caused by the rats that ate the palm seeds and by man himself who used the trees for fuel and for their oceangoing canoes. About this time, porpoise, which weigh up to 165 pounds, as well as tuna and other pelagic fish disappeared from their diet, as determined by examination of their middens. Starvation and wars between tribes followed. Without the trees and becoming overpopulated, the Rapa Nui culture collapsed. Adding to the civilization's demise, European explorers brought with them Western diseases like syphilis and smallpox. The Easter Islanders had no idea what the consequences were of their deforestation practices. I'm sure they had no malicious intent to self-destruct. If only they had had cameras they just might have had a record of the virginal island their ancestors first saw. In hindsight, they should have practiced rational thought and been environmentally aware, but obviously they didn't and weren't. [54]

Easter Island is not unique. Another example of humans altering their environment is New Zealand, which had been geographically isolated for 80 million years. During its long isolation, New Zealand developed a distinctive biodiversity of unique species of animals, fungi and plants. Because of its remoteness, its two islands were one of the last lands to be settled by Polynesians, between 1250 and 1300 AD. Before the arrival of humans in 1642, an estimated 80% of the land was covered in forest. Deforestation commenced shortly after their arrival. With the arrival of Europeans, logging and clear-cutting reduced the native vegetation to about 23%. Half of the native vertebrate species have become extinct; 51 species of birds, 3 frog species, 3 lizard species, 1 fresh water fish species and 1 bat species. [55, 56]

*Agroecology* is a whole systems approach to agriculture and food systems. It is based on an ecological approach that balances environmental soundness, social equity, and economic viability. Inherent in this definition is the idea that sustainability must be extended not only globally but indefinitely in time to all living organisms including humans. Adherents support protecting and restoring ecosystems and ending old-growth forest logging and fossil fuel dependency. Agroecology is a humanistic approach to agriculture. [57]

John Muir (1838-1914), the Scottish-American naturalist, petitioned the US Congress for the National Park bill that was passed in 1890, establishing Yosemite and Sequoia National Parks. In 1903 President Theodore Roosevelt accompanied Muir on a visit to Yosemite. After entering the park and seeing the magnificent splendor of the valley, the president asked Muir to show him the real Yosemite. Muir and Roosevelt set off largely by themselves and camped in the back country. The duo talked late into the night, slept in the brisk open air of Glacier Point and were dusted by a fresh snowfall in the morning. It was a night Roosevelt never forgot. Muir was able to convince Roosevelt that the best way to protect the valley was through federal control and management. The Sierra Club, which John Muir founded, is a prominent American conservation organization.

Too common is humans behavior to enrich themselves without regard to future consequences. Common is the notion that Nature is God's gift to mankind to exploit. In contrast to this sentiment, there are numerous environmental and conservation organizations dedicated to preservation of our natural heritage. There are intergovernmental organizations such as the Intergovernmental Panel on Climate Change (IPCC) and governmental agencies like the Environmental Protection Agency (EPA). There are about 150 US environmental organizations that are non-profit, non-governmental. An arbitrary small listing of these non-profits would include *The Center for Biological Diversity, Greenpeace, Nature Conservancy, The Ocean Conservancy,*

*Sierra Club, Rainforest Action Network, Global Coral Reef Alliance, Riverkeeper, Earthjustice, Earth First!* and *Natural Resources Defense Council (NRDC).* [58]

*NRDC* is a nonprofit international environmental advocacy group with a membership of more that 2 million members. They have a staff of nearly 500 lawyers, scientists and policy experts. NRDC's stated priorities include curbing global warming, reviving the world's oceans, defending endangered wildlife and wild places, and preventing pollution in order to ensure safe and sufficient water. Their motto is "The Earth's Best Defense." The NRDC is one of the most effective environmental advocacy organizations. They have been most successful by bringing lawsuits against polluters and lawsuits to prevent destructive governmental policies. The charity monitoring group *Charity Navigator* gave the Natural Resources Defense Council four out of four stars.

As an aside, a personal note: If you have concerns about humans future, consider supporting or endowing charities whose goals are "Saving the Earth". There are many worthwhile environmental causes and organizations. These organizations work to preserve our earth and our inhabitants. All these organizations rely upon member support. Support could be your legacy gift. I support the NRDC and the Sierra Club. Their advocacy agenda is aligned with mine and as such are a part of my legacy gifts.

## Behavior: *Over Populating*

*Not to have children is a selfish choice. Life rejuvenates and acquires energy when it multiplies: It is enriched, not impoverished!*

—Pope Francis, *Vatican Radio,* February 12, 2015

We are well over 7.3 billion people on this planet. [59] There will be about 128 million births this year and about 53 million deaths for a net gain of 75 million inhabitants on this planet, this year. Next year expect a bit more. We will be needing to build nine new cities every year, each the size of New York City. There is a finite amount of resources on earth. Any suggestions?

What I heard 50 years ago from one college sophomore, concerning over-population was: *We'll just head off into space and find another planet.* Consider that if each rocket ship holds 100 passengers we would need to launch over 2,000 of these rocket ships each day, Sundays included, just to keep pace.

In 1958, my professor at UCSB, Dr. Wally Muller, lectured about population growth. Then the world's population was about 2.8 billion and he thought that number of people was too great to sustain quality life. He noted that 10 years earlier, 1948, the population was 2.3 billion. He was alarmed that the world population had increased by 1/2 billion people in just 10 years, a 22% increase. We now have, in the intervening 58 years, increased the population nearly three times. I am certain Dr. Muller would be utterly shocked and dismayed of the present situation. Projections are that we will reach over 16 billion by century's end (if we don't kill ourselves off).

In 1798, the world's population was about one billion people. That is the year Thomas Malthus (1766-1834) wrote *An Essay on the Principle of Population.* Malthus was a British cleric and scholar in political economics and demography. His essay argued that population growth increases exponentially while the food supply increased arithmetically. The inevitable consequence is that people would outstrip available food, thus war, famine and disease are necessary checks on population. To forestall the dire consequences, he proposed preventive measures, either with moral restraints such as abstinence or delay of marriages or by restricting marriages of people suffering poverty or having defects.

What Malthus could not know is that modernization of farming would postpone temporarily the intersect of population and starvation. One of the most significant processes that changed agriculture was the discovery that ammonia could be synthesized from hydrogen and nitrogen. Ammonia is the basis of fertilizers. Fritz Haber (1868-1934), a German chemist, received the Nobel Prize in Chemistry in 1918 for his tabletop demonstration of the catalytic formation of ammonia from hydrogen and nitrogen under conditions of high temperature and pressure. Together with Carl Bosch (1874-1940),

Haber then developed an industrial process to produce fertilizers as well as explosives. Bosch also received a Nobel Laureate in Chemistry, in 1931. Prior to this time, there were two methods of increasing crop yields: natural fertilizers from animal wastes, dung, or alternatively farmers rotated crops with legumes, which fix nitrogen naturally.

*Unintended Consequences:* Ironically these two, by inventing the commercialized production of ammonia, increased the world's food supply and consequently the world's population by billions. They are also indirectly responsible for the destruction of millions of people, as ammonia is a constituent of explosives.

In 1968, when the earth's population had grown to about 3.5 billion, Paul Ehrlich (born 1932) published The Population Bomb. [60]

Paul Ehrlich, Professor of Biology at Stanford University, believed that nothing could be done; that the battle to feed all of humanity was over. He warned of the mass starvation of humans that would occur in the 1970s and 1980s due to overpopulation as well as other major societal upheavals. He advocated immediate action to limit population growth. Since the book's publication the world's population has doubled.

The major factor in this more recent population growth is *The Green Revolution.* Norman Borlaug (1914-2009) was an American agronomist, humanitarian and Nobel Laureate and is recognized as the *Father of the Green Revolution.* He was instrumental in the introduction of modern agricultural production techniques, hybridized seeds and synthetic fertilizers and pesticides to farmers of Mexico, Pakistan and India. He is credited with saving over one billion people from starvation.

*Feed and Breed* — words that rhyme and are linked.

During my senior year at UCSB, 1960-61, I took a year off from my studies and with my two friends, Jurgen Hilmer and Pete Rezendes traveled through Europe and across North Africa from Morocco to Egypt. We had an older VW van in which we slept, cooked and traveled, the classic 1960s adventure. In Egypt we traveled from Alexandria to Cairo to the Pyramids of Giza and on up the Nile towards Aswan. On the way up the Nile we visited a sugar cane refinery and met with the director of the refinery. During our visit we discussed the Aswan Dam, still under construction with aid from the USSR. He discussed the benefits that this dam would bring to Egypt. Damming the Nile would prevent the annual flooding. Also there could be year-round water available to irrigate greater acreage, hence a greater abundance of food. Unemployment was high and damming the Nile River would certainly improve living conditions for the Egyptians.

My thought at the time, and we discussed this, would not increased food production lead to increased population growth and hence the same standard of living? The Director sadly acknowledged that the population would increase. In 1961 the population of Egypt was 28 million essentially all living along a narrow strip of land bordering the Nile River. Today the population exceeds 90 million. They are still impoverished and political turmoil has increased.

In addition there were two unforeseen consequences of the Aswan Dam. The annual flood waters had brought nutrients down the Nile and into the Mediterranean Sea. The flooding ceased and, as a consequence, their shrimping industry. The nutrients were vital for the shrimp. The second consequence was that since there was year-round irrigation of fields there was a much higher incidence of schistosomiasis. Schistosomiasis is a parasitic disease caused by trematodes. These worms spend part of their life cycle in water snails and the other half in mammals, in the Egyptian case, humans. With increased flooded fields the population of snails increased and of course the incidence of this disease. Egyptian farmers are now wearing rubber boots and gloves when tending their fields. While the mortality rate of schistosomiasis is now low, there is damage to internal organs and, in children, impaired growth and cognitive development.

The United Nations Food and Agriculture Organization estimates that about 805 million of the 7.3 billion people in the world, or one in nine, suffered from chronic undernourishment in 2012-2014. Almost all the hungry people, 791 million, live in developing countries, representing 13.5% or one in eight of the population of developing counties. There are 11 million people undernourished in developed countries. Three billion humans depend upon the ocean for food. [61]

Garret Hardin, my UCSB professor, addressed the issue of human overpopulation. Dr. Hardin used a lifeboat as a metaphor to the "Spaceship Earth". There are limited resources. The lifeboat is near capacity. More humans may sink the boat. Who shall be permitted to board? The ethics of the situation stem from the dilemma of whether and in what circumstances should swimmers be taken aboard. Unfortunately, Spaceship Earth has no leader capable of enforcing any decision. [62]

So there! Once again humans, when faced with an impending disaster, rise to the occasion to stave off the disaster or more realistically we just postpone what will be a greater disaster. Feed the world and you merely increase the number of impoverished humans and they will breed. We're at 7.3 billion and counting.

The Catholic Church's stand on women's rights, abortion and birth control is archaic and not likely to change. The Church is ruled by men and that won't change. Humans continue to reproduce at rates greater than replacement. Girls and women need to be educated and provided with free birth control.

*Spike the Coke.* Create a great tasting soft drink, such as the popular Coca-Cola, but with the addition of a birth-control drug. Distribute it free or at a very low cost. I wonder what religious groups, pro-growth advocates and paranoid nationalists would say to such a suggestion. Realistically nothing can be done. Live with it and be prepared for the consequences. What more can be said?

## Behavior: *Energy Consumption*

*To truly transform our economy, protect our security, and save our planet from the ravages of climate change, we need to ultimately make clean, renewable energy the profitable kind of energy.*

—Barack Obama

For early man, a campfire was sufficient. In 2008 the world energy consumption was 143,851 TWh (terawatt-hour) = $10^{12}$ watts. About 81% of our energy comes from fossil fuels, 5% from nuclear and 16% from renewable. [60] Converting this number to a personal level you could use your share by burning 2,300 incandescent lightbulbs rated at 100 watts continuously for a year. Of course some of us use more watts, especially industry, and some of us use considerably fewer. By the year 2020 we will still obtain 80% of our energy from fossil fuels. [63]

If you have followed the ongoing debate concerning climate change, you know that environmentalists believe that we should switch over to alternative fuels - solar, geothermal, biofuels, wind power, hydroelectric, tides and waves. Continued usage of nuclear energy is debatable.

Let's be logical and rational and find a solution. As humans we come upon a problem, we solve the problem. Obviously we need energy. The solution is obvious. We will simply do away with carbon-based fuel. We'll rely upon renewable resources. Simple solution; problem solved.

How much energy are we using and what will be our future needs? How much of our future needs could renewable energy provide? What would be the downside if we were to make the switch? The huge human concern: what would be the long-term costs of *not* switching? Considering how difficult it is to convince not only legislators in Congress but the general public, is it possible to switch even to fluorescent lightbulbs? People are resistant to change.

*The answer my friend, is blowin' in the wind.*

—Bob Dylan

Present estimates of energy generated today from renewable energy sources range from 16 to 19% worldwide. Nuclear power generates another 10%. However, nuclear power energy production has fallen to 4.3% due to safety concerns and disasters. At best we are only about 1/5 of the way to becoming fossil fuel independent. The number of solar and wind generators is increasing, quite noticeably in Europe. The downside to switching is in my opinion rather small compared to the cost if we do nothing. There are a few minor concerns: locating wind farms near population centers presents problems, shadow-casting and noise. Bird collisions do happen, regardless of where they are located. Jobs lost from the petroleum industry will most likely be recovered by new green technologies.

Wind-power capacity has expanded rapidly to 336 GW (billion watts) in June 2014 and wind energy production was around 4% of total worldwide electricity usage and growing rapidly. [64] The top four wind-power producing countries in 2012 were the US at 140.9 TWh, China with 118.1 TWh, Spain at 49.1 TWh, and Germany with 46.0. Recall that 2008 World energy consumption in 2008 was 143,851 TWh. [65]

Can the switch be accomplished as long as elections are to a large part financed by fossil fuel corporations? Will the general populace embrace green technologies? Yes, as long as the switch will not affect them too greatly economically. There will also be some reluctance to go green by those few who always seem to believe that there is a conspiracy going on. An evil collaboration between the sinister climate scientists and, who knows, the CIA, Big Government and the Liberals. Maybe Barbra Streisand is the chief conspirator. If you were to know any scientists, you would realize that most are independent, analytical and skeptical; they couldn't possibly keep a conspiracy secret as they are always collaborating and sharing data. They are far too busy for such nonsense. Sorry, no evil conspiracy amongst climatologists.

"We can't take a carbon-free source of energy off the table," said Carol M. Browner, a former head of the *Environmental Protection Agency* who is now with *Nuclear Matters*, a pro-nuclear industry-backed group. Germany is planning to close all 17 of its reactors as part of an ambitious transition to renewable energy. "You can't rationally bet a big part of your climate change abatement plan on a technology that you may suddenly find you don't want to use anymore," stated Peter Bradford, a former member of the Nuclear Regulatory Commission. [66]

Natural gas now produces 27% of the electricity generated in the US. Burning natural gas produces about half the carbon dioxide that coal does for the same energy output. Unfortunately methane gas escapes into the atmosphere in the process of drilling and transportation. "The best role for natural gas is as a complement to renewable energy sources. The cost of solar has dropped 80% in the last five years, and the price of wind power has dropped by about half," said Hal Harvey, CEO, *Energy Innovation*, a company that develops strategies for clean energy.

Mr. Steven Cohen, the executive director of the *Earth Institute* at Columbia University, has taken a realistic view. "Everybody's looking for a big fix, a magic bullet. I don't think any of the existing technologies will do what we need to do to get to the renewable energy economy. The challenge is to spur the same kind of technology revolution that transformed bulky computers into pocket-size devices. You need some kind of technology that doesn't exist now. [67]

The petroleum industry won't keep quiet and for good reasons. The first reason is that they produce fossil fuel energy and have too much at stake to just abandon this extremely rich source of revenue. The petroleum industry creates millions of jobs, the executives make tons of money and the stockholders, most anyone who has a retirement plan, have a stake in the energy market. A sudden switch away from fossil fuels would create a worldwide crisis. The second equally important reason why switching is problematic is that these alternative sources of energy may not be sufficient to power our needs. Therein lies the conundrum. Is it possible to provide sufficient energy for the entire world's needs without the reliance upon fossil fuel?

Adam Frank in a *NY Times* op-ed asked the question: Is a climate disaster inevitable? "The defining feature of a technological civilization is the capacity to intensively harvest energy. But the basic physics of energy, heat and work known as thermodynamics tell us that waste, or what we physicists call entropy, must be generated and dumped back into the environment in the process. Human civilization currently harvests around 100 billion megawatt hours of energy each year and dumps 36 billion tons of carbon dioxide into the planetary system, which is why the atmosphere is holding more heat and the oceans are acidifying. As hard as it is for some to believe, we humans are now steering the planet, however poorly." [68]

There is a canary in the coal mine and he is not looking too happy. Let's attempt to cheer him up, just a little.

On September 6, 2012, Denmark launched the biggest wind turbine in the world and will add four more over the next four years. Additionally Denmark's infra-structure is well maintained. Most conservatives consider Denmark to be a socialist state. Most Danes are content to live in Denmark. [69]

2015, 42% of electricity in Denmark was generated by wind. *At 39.1%, last year's amount of wind-generated electricity was more than double what it was a decade ago. There was nothing extraordinary about the amount of wind in 2014 and the increase in electricity production can be attributed in part to the more than 100 new offshore windmills that were installed in 2014,* stated Denmark's Climate Minister Rasmus Helveg Petersen. Energinet is the Danish national transmission system operator for electricity and natural gas. It is an independent public enterprise owned by the Danish state under the Ministry of Climate and Energy.

Denmark aims to end the burning of fossil fuels in any form by 2050, not just in electricity production as some other countries hope to do, but in transportation as well. They are above a 40% level of renewable power on their electric grid, aiming at 50% by 2020. The political consensus to keep pushing is all but unanimous. Neighboring Germany, which has spent tens of billions pursuing wind and solar power, is likely to hit 30% renewable power on the electric grid in 2016.

After installation, wind and solar costs are minimal, which equates as large profits. That would drastically increase the benefit-to-cost ratio, rendering conventional power plants uneconomical to run. The potential problem is that any time there would be a blackout attributable to wind or solar, public support for the transition could weaken drastically. [70]

President Obama, on March 31, 2015, introduced a blueprint for cutting greenhouse gas emissions by nearly a third by 2025. This was part of a joint pledge made by Obama and President Xi Jinping of China. Republicans have called this blueprint a *war on coal* and an abuse of executive authority. Republican leaders vowed to weaken or undo it. [71]

*No one has ever become poor by giving.*

—Anne Frank *Diary of Anne Frank*

## Behavior: *Generosity*

There have been many studies that examine generosity. One study measured households in the US to determine which households were the most generous. My conservative friends are certain that it is the wealthy households making significant charitable gifts. In one sense they are correct but only in terms of the most money given. Most charitable gifts are tax-deductible. Comparing giving with their actual annual income, it is the individual in the lower income group that is the most generous. The *Independent Sector* is a nonprofit, non-partisan coalition of more than 700 national organizations, foundations and corporate philanthropy programs, collectively representing tens of thousands of charitable groups in every state across the nation. In 2001, an independent sector national survey found that those with a household income of less than $25,000 contributed 4.2%. Households with incomes in excess of $75,000 contributed 2.7%.

A more recent study conducted by the *Chronicle of Philanthropy* using tax-deduction data showed that households earning between $50,000 and $75,000 per year give an average of 7.6% of their discretionary income to charity. That compares to 4.2% for people who make $100,000 or more. In some of the wealthiest neighborhoods, with a large share of people making $200,000 or more a year, the average giving rate was 2.8%. [72]

Another study in 2011 found that Americans with earnings in the top 20% contributed 1.3% of their income to charity in comparison to those in the bottom 20%, who donated 3.2% of their income; that would be 2 1/2 times greater. The relative generosity of lower-income Americans is accentuated by the fact that, unlike middle-class and wealthy donors, most of them cannot take advantage of the charitable tax deduction because they do not itemize deductions on their income-tax returns. [73]

# Six

## Constructs

Human constructs are defined as complex ideas and theories that humans have invested resources, time and effort to uphold, not necessarily physical objects.

### Constructs: *Religion*

*When the missionaries came to Africa they had the Bible and we had the land. They said 'Let us pray'. We closed our eyes. When we opened them we had the Bible and they had the land.*

—Desmond Tutu (b.1931) South African social rights activist and retired Anglican bishop who rose to worldwide fame during the 1980s as an opponent of apartheid.

Religion is an ancient and complex human construct. Religion is most likely not a genetic trait. It is a coping mechanism. Religion offers explanations as to the meaning of life, death, hardship and an afterlife. Religious practitioners often experience comfort, piece of mind and occasionally exhilaration. The appeal of religion has persisted from very early times. Factors that can be attributed to the appeal of religion are numerous. Religion offers hope, serenity, security, conformity, codes for ethical behavior, explanations for the unexplainable and for the meaning of life. Religious practice must have arisen as an integral part of early tribal life. It provides a form of leadership and it creates a community with commonalities. It remains vibrant as a coping mechanism and as a form of tribal identity.

Religious congregations practice compassion with many of the unfortunate. Food and clothing banks are common endeavors of religious groups. There are numerous non-religious groups that serve the unfortunate as well. Goodwill, United Way and the Red Cross along with religious groups step up aid station in times of disasters. Compassion and empathy are very endearing human traits. What's disturbing is when religious dogma influences politics and when religious missionary groups link compassion with recruitment.

One of our ancient ancestors, a resourceful person, may have offered explanations for the unexplainable. That was the genesis of religion. Intelligent and creative individuals may have offered more elaborate explanations, a product of their imagination. Offering explanations most certainly would have generated a following. The emergence of a priest-like individual may have created a contentious situation with the leader of the tribe. These two positions, priest and chief, have continuously battled for supreme control over the tribe or congregation or nation from early times to the present. There is no record of the history of the emergence of religion as surely this must have happened thousand of years before recorded history.

Successful religions are usually well regulated. Questioning authority is often forbidden. Members rarely question the authenticity of the Bible. One never questions the Quran. Most all followers of faith accept their holy text as sacrosanct, accepting all within on faith. The fundamentalistic factions within each religion practice strict adherence to their ancient doctrines and ideology, or their preferred interpretations and revelations of those doctrines. There is a propensity for religions to send missionaries to obtain converts. Growth in numbers seems to be a necessity for some sort of satisfaction. Many conflicts and wars involve religious tribalism. The Crusades and the terrorist attacks in Paris are examples.

*Religion is what keeps the poor from murdering the rich.*

—Napoleon Bonaparte (1769-1821)

Religious leaders tell us about heaven. It is often stated that after death we will be together again with all our loved ones, at God's feet, if we're fortunate to have not sinned too greatly. There are many questions that are never addressed. What age are we all? Is Grandmother a young girl? Are we able to communicate with her? Would she know about our life after her passing? Would a baby that died always be a baby in heaven unable to communicate? Is heaven a beautiful garden with flowers always in bloom with butterflies fluttering about? Would there be any mosquitoes or other annoying insects? What about the 72 virgins and the eternal erection that Islamic *Martyrs* believe will be in their future? Why would religious leaders say these questions are unknowable or trite yet say of our deceased relatives they are in a better place. How could they possibly know?

Many athletes point up to the sky when entering the field or when at bat with the assumption that God will bless them with a home run or whatever. If players from opposing teams offer similar prayers for a favorable outcome, does God make a choice as to which player or team he prefers? We know that's ridiculous. It is often stated that God doesn't play favorites. Then why do athletes pray for positive outcomes? Occasionally your prayer is answered, or are they? That which affects the outcome can often be better understood by applying scientific principles. Religion is powerful and irrational.

The persistence of religion has some genetic components. Religion may produce feelings of euphoria. Beta-endorphin, an opioid, is found in the brain and gastro-intestinal tract. They are powerful analgesic or pain-killer molecules in humans. They influence brain activity by inhibiting the transmission of pain signals. Beta-endorphins can produce a feeling of euphoria. Religious beliefs and prayer introduce a profound sense of comfort in the face of adversity. This calm state may be partly due to the release of beta-endorphins. Ritualistic practices such as flagellation, fasting, dancing or other rhythmic movements including group singing in a church are particularly effective at stimulating the production of endorphins. These religious practices may have been designed to induce endorphin levels. [1]

Religion is often shrouded in mysticism. As a metaphor, consider the complexity of a modern automobile to that of religion. Look under the hood of a contemporary auto and you may not recognize any of the components; they are all covered with shrouds, similar to religion. A mechanic schooled in the internal combustion engine does understand how the engine components function and how they are integrated such that the car moves forward. He knows what is under the shrouds and why the car companies place shrouds - to protect the owner from sticking his hand in the fan or onto the battery. The shrouds may also be there to keep the owner from making their own repairs. Shrouds add a bit of mystery and provide a source of revenue for dealerships and for religions. Religion is similarly shrouded, a veil of mystery to prevent the practitioner from too closely examining the tenants of religion. Adam and Eve, Noah and the Ark, water into wine should not be examined. Fearfulness is necessary for continued adherence to religion. There is fear of the consequences of not being on the side of God.

*You don't need religion to have morals. If you can't determine right from wrong, then you lack empathy, not religion.*

—Anonymous

*I do not find in orthodox Christianity one redeeming feature.*

—Thomas Jefferson

The universe, the earth and all life that exists and has existed can be appreciated and understood without the need for any supernatural cause. Individuals that have "experienced" God or a supernatural energy may have not rigorously explored rational explanations for their experiences. Science cannot suggest a purpose for life nor answer questions concerning the afterlife. Science requires a great deal of study, understanding, background and intelligence. It is not readily accessible to the general public without intense study. All the "miracles" and cures of the past centuries were built to a great extent upon the science of western civilization. Science does not acknowledge either the presence or absence of a god or a creator. God was a concept created by man to give explanations to what is or was unexplainable. The Theory of Evolution has no moral or ethical principles and it makes no judgment of the atrocities of men. We readily accept theories concerning gravity, light, electromagnetic radiation, theories that help us understand electrons, nuclear power and how the body functions. Surprisingly, slightly more than 50% of residents of the US do not believe in evolution, rather they believe that evolution is contradicted by their Bible and that man just recently appeared on earth about 10,000 years ago as a creation of God. Dinosaurs are problematic. Faith is the suspension of reason. Spiritual enlightenment, the human soul, a vital force, energy fields - all leading to an evolved humanity living in global harmony, are all poetically compelling, but these New Age thoughts are well beyond the scope of scientific inquiry. Unfortunately some individuals believe that spirituality transcends science.

*It is forbidden to kill; therefore all murderers are punished unless they kill in large numbers and to the sound of trumpets.*

—Voltaire (1694-1778), French Enlightenment writer, historian, and philosopher famous for his wit, his attacks on the established Catholic Church and his advocacy of freedom of religion, freedom of expression and separation of church and state.

Philip Kitcher is a professor of philosophy at Columbia University. He expressed his views on religious doctrines. "To my mind, the experiences labeled religious come in two main types. There are some best understood in psychiatric terms. There are others, perhaps the overwhelming majority, that happen to people when they feel a great sense of uplift, often at the rightness of things. As Dewey pointed out, referring such experiences to some special aspect of reality is gratuitous speculation. Experiences of this sort are felt by completely secular people who classify them without appeal to religious language." [2]

A meta-analysis of 63 studies conducted by Dr. Miron Zuckerman of the University of Rochester showed a significant negative association between intelligence and religiosity. He discovered that intelligent children grasp religious ideas earlier, that they are also the first to doubt the truth of religion and that intelligent students are much less likely to accept orthodox beliefs and rather less likely to have pro-religious attitudes, concluding that perhaps intelligent people have less need on average for religious belief and practices. [3]

The sixth century BCE Greeks are credited with establishing the humanistic approach to humankind. The Greeks moved intellectual thought from myth-based religious explanations to physical explanations. Nature was not manipulated by arbitrary and willful gods or by blind chance. This is the beginning of scientific thought; theoretical thinking and systematization of knowledge. Greek physicians drew a distinction between medicine and magical powers, a radical shift for human civilization. Socrates (470/469-399 BCE) believed that ethics should be determined by rational thought, not from the realm of authority, tradition, dogma, superstition and myth. Reason was the only proper measure for questions of good and evil. Plato (c. 425-c. 348 BCE), Socrates' student, insisted on the existence of a higher world of reality independent of the world of things experienced every day. This higher realm of Ideas or Forms was unchanging, eternal, absolute, and has universal standards of beauty, goodness and justice. Truth resides in the world of Forms and not in the world revealed through the human senses. Aristotle (384-322 BCE) was the great appreciator of knowledge obtained by experiencing nature. He believed that to obtain knowledge of the world we must study the natural world through our senses. The Greeks could be credited with founding the rational and humanist tradition.

The Hellenistic Age(323 BCE-31 BCE), a Grecian-Roman period, expanded upon the rational traditions of early Greek philosophy. This age regarded the cosmos as governed by universal principles of order, *logos* or reason, accessible by rational thought. This school, called *Stoicism,* taught that virtue was the greatest good and is based upon knowledge, that human law should not conflict with natural law. Each individual was of one universal community, and rank was of no significance. The American Declaration of Independence is based upon the Stoic principle of inalienable rights. [4]

*I do not feel obliged to believe that the same*
*God who has endowed us with sense, reason and*
*intellect has intended us to forgo their use.*

—Galileo Galilei

*The Age of Enlightenment* can be traced to 17th and 18th century Europe and on to the American colonies. It was a challenge to faith, tradition, superstition and the intolerance and abuses of power by the church and state. Its purpose was to reform society by reason and the scientific method. Early proponents were Spinoza, John Locke, Voltaire and Isaac Newton. *The Age of Enlightenment* greatly influenced American colonists Ben Franklin and Thomas Jefferson, as reflected in the Declaration of Independence and the Bill of Rights. [4]

## Constructs: *Politics*

*One of the penalties for refusing to participate in politics*
*is that you end up being governed by your inferiors.*

—Plato

*Politics is a highly tribal business.*

—Nick Clegg (b. 1967), British politician who was
Deputy Prime Minister of the United Kingdom.

Politics is defined as of, for or relating to citizens, the practice and theory of influencing other people. It is a framework that defines acceptable political methods within a society. More narrowly it refers to achieving and exercising positions of governance. The word *Politics* can be traced back to the works of Confucius, 551–479 BCE, and to the fourth century BCE Greek philosophers in such works as Plato's *Republic* and Aristotle's *Politics.*

Democracy is a system of government by the whole population or all the eligible members of a state, typically through elected representatives. The function of most governments is to provide their citizens with protection from foreign intrusions as well as governance of their inhabitants, which would include regulating economic policies to various degrees. Origins of democracy can be traced to the 5th century BCE. "Rule of the people" denoted the political systems that existed in Greek city-states, notably Athens. Democracy is the antonym of "rule of an elite". Democracy is not an economic system; however, politics and economic policies are intertwined.

*With law shall our land be built up, but*
*with lawlessness laid waste.*

—NJAL'S SAGA, ICELAND, C1270. THE SAGA DEALS WITH THE PROCESS OF BLOOD FEUDS IN THE ICELANDIC COMMONWEALTH, SHOWING HOW THE REQUIREMENTS OF HONOR COULD LEAD TO MINOR SLIGHTS SPIRALING INTO DESTRUCTIVE AND PROLONGED BLOODSHED.

Legal systems began at a time when pastoral life was being replaced by life in villages, towns and cities. One of the oldest legal documents that we know of is a code of laws enacted by the Babylonian king Hammurabi in about 1754 BCE. The Code consists of 282 laws with scaled punishments governing household behavior, marriage, divorce, paternity, inheritance and payments for services. Included is "an eye for an eye, a tooth for a tooth" and graded depending on one's social status, slave versus free man. An ancient 2.24 meter stele inscribed with Hammurabi's Code can be seen in the Louvre. The laws are written in the Akkadian language using cuneiform script. [5]

Politics, as practiced today, is a broad spectrum of ideas, doctrines and values. In the US as in many countries, there is a great variety of politics that range from conservative to liberal values. For the most part, citizens are about equally divided into these two categories. We are all affected by politics and we are involved in politics either actively or passively. Few are dispassionate. Most belong to or feel associated with a political group. It's as if we belong to a tribe.

Television is one source of world and national news. Many of you may remember Walter Cronkite (1916-2009), anchorman for the *CBS Evening News* for 19 years. One could expect this evening news broadcast to cover timely events worldwide without an obvious political slant. Not so today. The news department of most "News" programs today are slanted towards their demographic, their tribe. The world is barely covered and a large segment, always at the end of the broadcast, is devoted to a heart-felt story that leaves us all teary-eyed, often a story featuring the accomplishments of some individual with a handicap overcoming adversity.

Obtaining timely news and information necessary to judge the politics of any country is difficult. The Internet is another source. All the major news organizations have an Internet presence that is nearly identical to their TV or print programming. Independent of these major outlets are bloggers with timely information that may be sourced or unsourced and fact-checked or not. The degree of editorializing varies. Perhaps it is better to just stick with your tribe's position and not waste time doing your own research.

During election cycles, the tribes compete to be heard; negative political ads, talk radio and TV personality rants. Newspapers show their bias by publishing their own editorials along with appropriate political *think tank* position papers. All these media are competing for the public's attention. The voices often offer incomplete and/or dishonest information. It's apparent that a great deal is at stake for each tribe.

*Think Tanks* or *Advocacy Groups* are devoted to articulating political and economic views; creating talking points, influencing public policy. These think tanks are large and very well funded and have enormous influence. Examples of these groups:

*The Heritage Foundation* is an American conservative think tank. Their stated mission is formulate and promote conservative public policies.

*The Cato Institute* is an American libertarian think tank founded by Charles Koch.

*The Center for American Progress* is a progressive public policy research and advocacy organization. It presents the liberal perspective.

*The Brookings Institution* conducts research in the social sciences, governance and domestic and global economic policies. It is politically independent.

> "*Capitalism is a big problem, because with capitalism you're just going to keep buying and selling things until there's nothing else to buy and sell, which means gobbling up the planet.*"
>
> —Alice Walker (b. 1944), American author and activist. She wrote the critically acclaimed novel The Color Purple.

Capitalism and socialism are both economic systems. While there are various models of Capitalism, it commonly refers to the economic system in which a country's trade and industry are controlled by private owners and goods and services produced are sold for profit in a market economy. The objective of capitalistic corporations is to survive, prosper and reward investors. Capitalism is not bound by democratic principles nor long-term consequences. Capitalist systems range from laissez-faire with minimal government regulation to having various governmental relationships such at special tax breaks and incentives. Capitalism can be practiced with social concerns, an alternative to pure capitalism.

The history of Capitalism has been traced back to the ninth century Islamic world. Sharia, or Islamic, law deals with many moral and religious topics as well as systems of politics and economics. Sharia encourages charity and discourages usury on loans.

A merchant form of capitalism started in 12th-century Europe and by the sixteenth century was a recognizable form active in The Netherlands, with the beginnings of full-time stock exchanges. Remembered from this era was the speculative economic tulip market. From November 12, 1636 to February 3, 1637, the price of some single tulip bulbs soared to ten times the annual wage of a skilled worker. On May 1 the same year, the market collapsed. This *Tulip Mania* is generally considered the first recorded speculative bubble.

Europeans capitalized on this early form of capitalism by building large ships to bring back treasures from the new world, Africa, Asia and most anywhere where they traded their religion and manufactured goods for new resources. Vast intercontinental trading brought great wealth to the seagoing merchants.

Greater wealth, in part, brought on *The Industrial Revolution.* The industrial leaders increased their wealth while the working class was at the mercy of those controlling the resources. What emerged was a three-class system: a very small exclusive wealthy ruling class, the merchant class and the large uneducated working class. Wealth equals power, and the separation between classes increased steadily. There are many individuals who would consider capitalism good, a system that promotes freedom and the good life.

Today in our capitalistic system of commerce, corporations are consolidating, in essence eliminating their competition and shipping their manufacturing overseas to lower their cost of production. Jobs are eliminated. Automation has also caused an increase in local unemployment as well as increased profits for corporations. Large stockholders and corporate executives, the top 1%, are joyful as profits soar and their dividends increase. The jobless can always go on "Family Assistance", that is until their benefits run out. Capitalism is based on perpetual growth, with the inevitable depletion of the world's natural resources. Long term it is unsustainable. Did humans adopt a better way of life with capitalism? A few certainly did.

*The DNA of a corporation has only two purposes: to solve its own problems and to help its shareholders.*

—Kevin O'Lear (b. 1954), Canadian entrepreneur, investor, journalist, writer, financial commentator and television personality. What he did not say: *Do no harm.* In January 2014, O'Leary reported on his television show, *The Lang and O'Leary Exchange,* that the news that the world's richest 85 people owned as much wealth as the bottom half of the global population and called it «*fantastic news.*"[6]

*Regarding the current political debate over government tax and spending issues, there are several points on which all conservatives agree. First, the federal government has grown way too large and is engaged in too many activities that transcend constitutional boundaries. Second, in this unrestrained role, the government is spending way too much money and running up way too large deficits. Third, this overspending and deficit irresponsibility threaten not only the very integrity and authority of government, but the foundation and health of the American economy.*

—Patrick Garry (b. 1955), Professor of law at the University of South Dakota and the Director of the Hagemann Center for Legal & Public Policy Research. He is the author of Limited Government and the Bill of Rights, 2012.[7]

In current politics, there is an ongoing battle between conservatives and liberals. The conservative trait could be characterized as having a self-centered nature, always looking out for themselves, limiting government, looking for the bargains, tax avoidance and accumulating wealth. Conservatives are sober, modest and cautious. Conservatives would argue that America would be a better and freer nation if we were to follow conservative principles. Conservatives champion capitalism and argue that fewer governmental restrictions would stimulate creativity and innovation and encourage economic growth that otherwise is hampered by governmental restrictions and oversight. The *Affordable Care Act* is too costly and unnecessary. Governmental food stamp programs encourage dependency and destroy two-parent families. These programs should be the provenance of private non-profits and religious groups. Governmental departments that regulate and police many activities such as food inspection should be self- regulated by the corporations, outsourced or eliminated. Governmental programs impede our individual freedom, and efforts by the liberals to level the playing field are a socialist money grab, a redistribution of wealth at the expense of the hard-working wage earners. These governmental programs are social engineering.

Cost-cutting measures that benefit corporations and their investors short-term rarely consider the long-term costs and benefits. Over regulation surely stifles innovation. All systems require governors. Any biological, societal or mechanical system without governance will self-destruct. There always seems to be an ongoing battle between the schemers and scammers and the regulators. When competition between corporations is fierce, with few exceptions, there will be shenanigans. Regulations are devised to alleviate this. Conservatives are worried that we overspend, overprotect, over regulate, overthink, overanalyze and that there are too many leftwing liberals, intellectuals, artists, poets and free thinkers destroying the American Way of Life. They abhor socialism and contend that it is forced or involuntary cooperation.

Some Conservatives certainly do favor certain laws: drug enforcement, mandatory sentencing, Defense of Marriage laws, laws that dictate morality and especially voter-fraud laws that have resulted in one dozen actual prosecuted cases over the last decade, nationwide. They do favor some government spending for more prisons but not higher education; let the students pay for their education. If their parents aren't wealthy, the banks offer student loans that the student can pay back with interest. Student-loan debt has topped credit-card debt as the largest form of consumer debt across the nation at $1.2

trillion and growing. Potentially adding to that debt, on July 1, 2013, student borrowers of new subsidized federal Stafford loans saw their rates double from 3.4% to 6.8%. [8]

Conservatives can at times be compassionate. Viewing the society pages of newspapers, one sees photographs of galas and society balls with the wealthy socialites dressed in splendid evening attire. Most certainly, these events do benefit worthwhile charities.

Dacher Keltner, a professor of psychology at Berkeley, "found that the poor, compared with the wealthy, have keenly attuned interpersonal attention in all directions. Generally, those with the most power in society seem to pay particularly little attention to those with the least power. In politics, readily dismissing inconvenient people can easily extend to dismissing inconvenient truths about them. The insistence by some House Republicans in Congress on cutting financing for food stamps and impeding the implementation of Obamacare, may stem in part from the empathy gap." [9]

*Compassionate Conservatism* was the campaign slogan of George W. Bush. The basis tenets, the core values of the conservatives position, are self-interest, pay little or no tax and have limited government. Limiting government primarily to defense is a core value for two reasons. It's a whole lot cheaper if we were to eliminate many governmental departments and agencies such as the Environmental Protection Agency, the Department of Education and the National Endowment for the Arts. These agencies and departments have been highlighted for elimination by conservatives, the EPA to eliminate regulations protecting the environment, and the Department of Education, for restricting religious teachings in public schools.

*The only difference between compassionate conservatism and conservatism is that under compassionate conservatism they tell you they're not going to help you but they're really sorry about it.*

—Tony Blair (b.1953) British Labour Party politician, who served as the Prime Minister of the United Kingdom from 1997 to 2007.

What should be the role of government? Remember what Ronald Reagan said: "The worst nine words you may hear is: I'm from the government and I'm here to help."

Really? Consider these 48 words: *Due to Federal budget cuts, Head Start and Food Stamp programs, The Center for Disease Control, food safety agencies, the Departments of Health and Human Services, Education, the EPA, NEA, NSF are dissolved. However, congressional members and their staff as well as the Department of Defense are unaffected.*

My best nine words you might or might not hear: *One hundred Wall Street manipulators have gone to jail.*

Is government too big? How big is too big? What Ronald Reagan implied is that outsourcing is better than having government providing service,; that the government should not provide social services. He believed that government was needed for defense, but not much more. Outsourcing does provide services for the government and has done so for decades. However, there are problems. Governmental contracts usually go to the lowest bidder or to corporations with the largest lobbyist funding. That does not necessarily mean there will be savings. To win the bid, corporations either need to be extremely efficient or take shortcuts that might be costing the government more than expected to correct defects. Oversight of many corporate projects is self-regulated or outsourced to private contractors.

The *Sunlight Foundation,* a watchdog group advocating for government transparency, analyzed the 200 largest US companies ranked by 2010 pre-tax income and found those that spent the most on lobbying between 2007 and 2009 had lower 2010 tax rates than what they paid in 2007. These 200 of America's most politically active corporations spent a combined $5.8 billion on federal lobbying and campaign contributions. A year-long analysis by the Sunlight Foundation suggests that what they gave pales compared to what those same corporations got: $4.4 trillion in federal business and support. That figure is more than the $4.3 trillion the federal government paid the nation's 50 million Social Security recipients over the same period. *After examining 14 million records, including data on campaign contributions, lobbying expenditures, federal budget allocations and spending, we found that, on average,* *for every dollar spent on influencing politics, the nation's most politically active corporations received $760 from the government. The $4.4 trillion total represents two-thirds of the $6.5 trillion that individual taxpayers paid into the federal treasury.* The top eight firms that paid the most in lobbying, resulting in lower tax rates are: ExxonMobil, Verizon, General Electric, AT&T, Altria, Amgen, Northrop Grumman and Boeing. [10]

The healthcare industry in 2014 spent $500 million seeking to advance policies or avoid setbacks on issues crucial to their interests. Have you noticed the big drop in drug prices and medical costs? Have you noticed the fancy cars and homes owned by the executives of the healthcare and insurance industries? Hard to notice, they all live within gated communities. [11]

One persistent attitude attributed to capitalism is that this form of commerce promotes motivation, the driving force for a vibrant economy. "Socialism is unsustainable as it destroys the motivation to work". This is a bit of a myth. Most people do want to work. Few want to live on the dole. What everyone agrees on is that with innovation, new and improved goods come to market. Rewards are given to the innovators and producers. The economy grows. The downside to this is that often more jobs are lost than gained. Competition between corporations becomes fierce, often with the industry laggers resorting to corrupt practices such as dumping waste into a water source as a cheaper method of disposal. When industry leaders profits increase, we often see their profits funneled to Washington lobbyists "asking" legislators to decrease their client's taxes. As profits increase, lobbying expenses follow.

Socialism, as well, has various models. One model, *Democratic Socialism* resonates today amongst liberals. Democratic socialism has concerns for social justice within the marketplace. Each individual should live a meaningful life with the full development of his or her personality and talents. It embraces a community in which power, wealth, and opportunity are in the hands of the many, not the few. Democratic Socialism argues that markets should be governmentally regulated to balance the inequalities that emerge from the market. Trade unions are integral to achieving their goals. For many, socialism is thought, mistakenly, to be incompatible with democracy.

Democratic Socialism is not to be confused with Communism, a socio-economic system structured upon the common ownership of the means of production.

*Capitalism is the worst enemy of humanity.*
*Capitalism and the senseless development of unlimited industrialization are what destroy the environment.*

—Evo Morales, President of Bolivia

Evo Morales, the first indigenous President of Bolivia, was elected to office in 2005 with 53% of the vote. Morales presides over a socialist economic system, a system of production and distribution organized to directly satisfy economic demands and human needs so that goods and services are produced directly for use, instead of for private profit driven by the accumulation of capital. Rather than trying to "live better," he said, our goal should be to "live well." "Capitalism has created a civilization that is wasteful, consumerist, exclusive, clientelist, a generator of opulence and misery. That is the pattern of life, production and consumption that we urgently need to transform." In 2009, Morales was reelected with a landslide majority, polling 64%. He was again reelected in the 2014 general election. [12]

Greater detail about capitalism in Bolivia and elsewhere can be found in Naomi Klein's excellent book The Shock Doctrine: The Rise of Disaster Capitalism, 2007, Metropolitan Books

José Mujica (born 1935), the President of Uruguay between 2010 and 2015, has been described as *the world's humblest president* due to his austere lifestyle and his donation of around 90% of his $12,000 monthly salary to charities that benefit poor people and small entrepreneurs. He still resides on the outskirts of Montevideo, cultivating chrysanthemums for sale. He declined to live in the presidential palace. His aged Volkswagen Beetle is his source of transportation. Asked about his frugal lifestyle, Mujica said, "I don't want to use the word 'austerity,' because they prostituted it in Europe", referring to controversial policies of deeply cutting government spending to confront some European countries' fiscal crises. "I live with little, with moderation, so that I can occupy myself with what's important. Basic humanistic values should transcend economic concerns." [13]

By numerous polls Denmark has been rated as the happiest country in the world, where citizens are the most content. Denmark, along with other Scandinavian countries, has high taxation, but with medical and educational benefits well beyond ours. Denmark's overall tax burden (sum of all taxes, as a percentage of GDP) is estimated to be 46% in 2011. By comparison the OECD countries (*Organisation for Economic Co-operation and Development)* overall tax burden is 36.2% and the USA percentage is 27.3%. [14]

Denmark has the world's lowest level of income inequality according to the *World Bank,* which also ranks Denmark as the easiest place in Europe to do business. Denmark also has the world's highest minimum wage as well as one of the world's highest per capita incomes. It also has the fourth highest ratio of individuals with advanced degrees. All college and university education in Denmark is free of charges. Along with Sweden and Norway, Denmark has a universal healthcare system financed by taxes and not by social contributions. [15]

Many New Testament scholars say that Jesus Christ was a socialist; he performed acts that most would identify as socialistic. He healed the sick and fed the poor presumably without renumeration. Today many would consider that Socialism is a very nasty word. We fling it around as we did the word Communist 60 years ago. Of the many epithets used to describe President Obama is that he is a Socialist, but not in a flattering manner. Walmart is a socialistic company. It is successful, in part, because, while paying its workers just a bit above minimum wage, Walmart relies upon the government to make up the difference in wages with food stamp and family assistance programs so that their employees can survive. Without governmental handouts, where would Walmart be?

*I believe in a relatively equal society, supported by institutions that limit extremes of wealth and poverty. I believe in democracy, civil liberties, and the rule of law. That makes me a liberal, and I'm proud of it.*

—Paul Krugman (b. 1953), American economist, Distinguished Professor of Economics at the Graduate Center of the City University of New York, and an op-ed columnist for *The New York Times*

The Democratic Socialism or Liberal agenda can be summarized by past accomplishments:

Medicare
Social Security
Securities and Exchange Commission
Federal Reserve
Food and Drug Administration
National Park Service
Federal School Lunch Program
Food Stamps
Voting Rights Act
Equal Access to Public Accommodations
Collective Bargaining
Fair Labor Standards
Clean Air Act
Clean Water Act
Equal Rights for Women

*Every one of these efforts, which strengthened our democracy and the quality of life in America, began as a liberal initiative contested by conservatives.*

—George McGovern [16]

A few of President Barack Obama accomplishments that benefited all Americans:

American Recovery and Reinvestment Act, 2009
Hate Crimes Prevention Act, 2009
Public Lands Management Act, 2009
Lilly Ledbetter Fair Pay Act, 2009
Affordable Care Act, 2010
Dodd-Frank Wall Street Reform and Consumer Protection Act, 2010
FDA Food Safety Modernization Act, 2011
Paris Climate Agreement, 2015

These accomplishments were also liberal initiatives contested by conservatives.

Martin Gilens, Professor of Politics at Princeton University and Benjamine Page, Professor of Decision Making at Northwestern University, have studied the degree of influence over public policy different groups have on the average citizen, individually in organizations, or on the economic elites and business groups. They performed multiple-variant analyses measuring key variables for 1779 US policy issues. Analyses have shown that the *economic elites* and *business interests* have the greatest substantial impact on US policies. This analysis substantiates the notion that the average citizen or mass-based interest groups have little or no impact on U. S. policies. The results have provided support for the theories of Economic-Elite Domination and Biased Pluralism and no support for the theories of Majoritarian Electoral Democracy or Majoritarian Pluralism. [17]

*The forces of a capitalist society, if left unchecked, tend to make the rich richer and the poor poorer.*

—Jawaharlal Nehru (1889-1964), first Prime Minister of India and a central figure in Indian politics for much of the twentieth century. He emerged as the paramount leader of the Indian independence movement under

THE TUTELAGE OF MAHATMA GANDHI. NEHRU IS CONSIDERED TO BE THE ARCHITECT OF THE MODERN INDIAN NATION-STATE: A SOVEREIGN, SOCIALIST, SECULAR, AND DEMOCRATIC REPUBLIC.

"One of the consequences of conservative economic practices is the disparity of wealth. Today, with a USA total wealth of 54 trillion dollars, the top 1% of Americans have 40% of the wealth and own 50% of all stocks, bonds and mutual funds. The bottom 80% have 7% of the wealth. The bottom 50% of Americans own 0.5% of stocks, bonds and mutual funds; they are just scrapping by. The Trickle Down theory in practice gives tax breaks to the wealthy since they are the "Job Creators". They will create jobs and we all will be better off. Here are some facts. In 1995, the Millionaires' Effective Tax Rate was 30.4%. In 2009, with the Bush tax cuts in effect, the millionaires' tax rate dropped to 22.4%. Unemployment in 1995 was 5.6%. However, the 14 years of lower tax rates for the rich resulted in an increase in unemployment to 9.3%. Obviously there were many factors to the joblessness, but tax breaks for the wealthy did result in greater unemployment, the exact opposite of what the Republicans promised." This quotation is by Nick Hanauer (b 1959), an entrepreneur and venture capitalist. Nick Hanauer's March 1, 2012 *TED TALK* was *not* posted on the *TED Talks* website. It is rumored that *TED* banned the talk as, *too politically controversial, that TED did not want to offend the wealthy attendees.*

In 1950 executives were paid about 20 times the wages of their workers. Now the average executive receives 240 times that of their average workers. Wages for the average worker have barely increased compared with executive compensation, which has skyrocketed. [18] Perhaps today's CEOs are working 12 times more hours per day than they did 60 years ago. That equates to working 96 hours per day. [18]

Oxfam has recently reported that the world's 85 richest individuals have the same wealth as 3.5 billion of the poorest. Translation: 0.0000012% of the earth's population controls 50% of the total wealth. The wealth of the richest 1% in the world amounts to $110 trillion or 65 times as much as the poorest 50% of the world. Oxfam also argues that this is no accident, saying growing inequality has been driven by a "power grab" by wealthy elites who have co-opted the political process to rig the rules of the economic system in their favor. [19]

In November 2013, the *World Economic Forum* released its *Outlook on the Global Agenda 2014.* Ranked as the number one risk are is the rising societal tensions in the Middle East and North Africa. The second worldwide risk is widening income disparitiy, impacting social stability within countries and threatening security on a global scale. [20]

In September 2013 a UC Berkeley study found that between 2009 to 2012 the average real income per family in the US for the top 1% grew by 31.4%. The bottom 99% of incomes grew by only 0.4%. The top 1% captured 95% of the income gains for these three years. [21]

When 1% of Americans own 40% of US wealth and with 442 billionaires in the US, can there be a true democracy?

*Citizens United v. Federal Commissions was* a constitutional law case in which the US Supreme Court held that the First Amendment prohibits the government from restricting political independent expenditures by corporations, associations or labor unions. Included in the majority opinion: *The appearance of influence or access, furthermore, will not cause the electorate to lose faith in our democracy.*

"Big donors contribute to politicians; not necessarily for ideological considerations, but rather for access; economics and self interest. If one accepts this assumption, then the one man-one vote concept is out the door. Can there be electoral integrity? Corporations have in fact replaced the legislative branch of government. They write the laws. They have caused University tuition to be accessible only to the ultra rich, or gamblers certain of their financial future." Jill Lepore (b. 1966) Professor of American history at Harvard University and chair of Harvard's History and Literature Program. [22]

The jobs of the future are either in the service industry or will require a high level of training and education. With global warming increasing there will most certainly be a future need for greater numbers of wildfire fighters and disaster relief workers. Service jobs are often at wages insufficient to support a family. Family assistance programs are being cut to bring down the national debt. High-paying jobs, those requiring a college or university education, require substantial costs and are mostly unavailable to those who do not come from rich families or not willing to take on substantial debt. Bank of America is doing quite well issuing student loans. Students with these loans will be working for the banks for decades. In the late 1950s, when I attended UC Santa Barbara, I was able to finance my education by part-time and summer jobs. I received no family financial assistance. I did not take a student loan. California's wise leaders, *pre*-Ronald Reagan, realized the need for educated people for California's future. Tuition fees were nonexistent, *then,* for residents. Anyone with the desire and qualifications could attend the university. With part-time work I managed just fine. Today the cost of tuition and attendance for California residents is $34, 500 per year, $138,000 for four years. For non-residents add another $24,000 per year.

*Taxes are the price we pay for a civilized society.*

—Oliver Wendell Holmes (1841-1935), American jurist who served as an Associate Justice on the US Supreme Court from 1902 to 1932.

How low Are US taxes compared to other countries? The US comes in 55th out of 114 countries in a graph from the Tax Policy Center of the Organization for Economic Co-operation and Development (OECD). We're at the bottom of the stack, 25% below the average. Denmark has the highest at about 47%, the OECD average is at 36.2%, the USA is at 27.3% and Mexico is the lowest at about 22%. [23]

*The Institute on Taxation and Economic Policy*, a nonprofit, nonpartisan research organization, published a study January 14, 2015, that stated: "The tax systems are upside down, with the poor paying more and the rich paying less. Overall, the poorest 20% of Americans paid an average of 10.9% of their income in state and local taxes and the middle 20% of Americans paid 9.4%. The top 1%, meanwhile, pay only 5.4% of their income to state and local taxes. If you need more evidence, consider modern Germany, where taxes on the wealthy are much higher than they are here and the distribution of income is far more equal. But Germany's average annual growth has been faster than that in the United States."

Tax avoidance is a legal method to reduce the amount of tax one owes, such as claiming the permissible deductions and credits, contributing to employer-sponsored retirement plans and or purchasing municipal bonds. Tax havens are jurisdictions, states or countries where some taxes are levied at a low or zero rate. Some corporations and individuals establish shell subsidiaries in tax havens as a tax-avoidance strategy. Tax evasion is the term for efforts by individuals, corporations and trusts to evade taxes by illegal means.

Tax avoidance may be legal, but is it ethical? A patriot is an individual who is devoted to and vigorously supports their country. Is it patriotic to avoid taxes? We feel it is the responsibility of government to provide services that are essential for this country - roads and bridges, the military, food safety, literally hundreds of services. The services are paid for by taxes. How patriotic are individuals who avoid taxes, even legally? Do the tax avoiders simply lack empathy for their country every April 15?

*Taxes should hurt.*

—RONALD REAGAN, 1971, ALTHOUGH HE AVOIDED PAYING STATE OF CALIFORNIA INCOME TAX. [24]

Is there a genetic link to one's political tendencies? It is widely assumed that social behaviors and political preferences are culturally transmitted. This assumption has been challenged by numerous studies of adult twins and their relatives, adoption studies and twins reared apart. One study with a different approach consisted of 13,201 individuals from 2,774 families whose DNA was analyzed and compared to their responses to a 50-item sociopolitical attitude questionnaire that gauged their Conservative-Liberal orientations. The focus of the research was to identify genetic focal points that may show correlation with political preferences. Genetic focal points or linkage regions of chromosomes are segments of any chromosome that can be chemically dissected, segregated and analyzed. These regions may contain numerous genes. This analysis compared linkage regions of the twenty-two human autosomal chromosomes (the sex chromosomes were excluded). On chromosomes 2, 4, 6 and 9, there were numerous linkage regions that correlated with political preferences. Several linkage groups showed correspondence of neurotransmitters and associated biochemicals, such as serotonin receptors, to political preferences. These researchers suggest that some genetic loci may influence flexibility in information processing and cognition as well as loci that regulate fear, anxiety, mate selection and disgust (which is an olfactory trait). Any pathway from DNA to social behavior is certain to be convoluted, involving networks of genes, genetic expression, multiple intervening neurobiological processes, development and a multitude of environmental contingencies. Conservatism and Liberalism have much to do with flexibility of opinion. The data gave preliminary support to the hypothesis that whatever relationship exists between politics and genetics it may be genetic loci that influence flexibility in information processing and cognition. The researchers readily admit that no gene is acting directly to determine our political views: there is no Liberal or Conservative gene, but there might be a combination of genes acting together that somehow predispose us to have particular politics, presumably through their role in influencing our brains and thus our personalities or social behaviors. [25]

People may know what is the accepted scientific view, but they are unwilling to believe it when it contradicts their political or religious views. Once people's cultural and political views get tied up in their factual beliefs, it is very difficult to undo, regardless of the messaging that is used. Unfortunately, knowing what scientists think is ultimately no substitute for actually believing it. More people know what scientists think about high-profile scientific controversies than polls suggest, they just aren't willing to endorse the consensus when it contradicts their political or religious views. [26]

However these is an abundance of literature that offers suggestions as to how to change people's positions. Howard Gardner of the Harvard Business School, has offered ideas mostly applicable for the business world: "Present promising idea with enough frequency and variety that others will understand it, remember it, and, most important, embrace it." Dr. Gardner suggests the following: Say it often and in many ways in a variety of formats. Frame your message objectively in neutral and familiar terms. Provide contrasting scenarios, a devil's advocate perspective. Importantly, know your audience's intelligence. Select the right blend of descriptions and representations. [27]

Susan K. Perry, Ph.D., a social psychologist suggests *5 Ways to Change Someone's Mind.* Keep the message simple. Connect to their best interest. Present some surprising feature or attribute. Show confidence in your point of view. Display empathy. [28]

There you have it. We now know how to change someone's political views. Sounds easy. Maybe too easy, as in *too good to be true.*

I am not completely disillusioned by humans. The unexpected sometimes occurs. During my years as a student at John Marshall Junior High School in Pasadena, CA there were annual Student Body Presidency elections. The candidates were offered a forum where each candidate spoke, presenting his or her ideas for a more perfect student body. I remember the two individuals running for Class President. Cathy was a bright normal-looking teenage girl. Her opponent was a popular handsome football player. I don't recall the football player's name, but I do remember Cathy. Her pledge was to increase student support for all the usual reasons and so forth. I do not remember exactly any of her pledges except that they all seemed reasonable and admirable. Her speech was warmly received. Next to speak was the football player. His speech as short. He pledged to replace the water in the school's water fountains with Coca-Cola. The students in our auditorium were cheering. Everyone thought that this was a great idea, free Coke. Everyone thought that the election would be a landslide. After a few days of campaigning, with the hallways covered with handmade posters, it was time to vote. It was a landslide. Cathy won. Rationality prevailed even for these young people. These young teenagers, given a bit of time to reflect and think about the feasibility and implications decided that "Free Coke" was just a cheap gambit without merit.

The future will be in the hands of the youth. It is to be hoped that there will be sufficient numbers of young people who will be educated and motivated to take up the challenge. There are scholastic activities that attract the youth: Science Fairs, Spelling Bees and other academic competitions that motivate the young to be champions. Unfortunately high school football games are better attended.

Why don't we: *Live Together in Perfect Harmony*

—PAUL MCCARTNEY AND STEVIE WONDER, *EBONY AND IVORY* (1982)

## Constructs: *The Sciences*

*Science is a very human form of knowledge.*

—JACOB BRONOWSKI

*We don't run the government, the government runs us, We live in the age of science and technology and few politicians have the education to understand either. Science is more than a body of knowledge, it's a way of thinking.*

—CARL SAGAN, INTERVIEWED BY CHARLIE ROSE, 1997

Science is a body of empirical, theoretical and practical knowledge about the natural world. Scientists use the power of observation and rational reasoning to predict and to explain the real world. Science is one of the pinnacles of the human experience. It is the expression of the creativity of humans.

*An acorn is potentially, but not actually, an oak tree. In becoming an oak tree, it becomes actually what it originally was only potentially. This change thus involves passage from potentiality to actuality—not from nonbeing to being but from one kind or degree to being another.*

—Aristotle (384-322 BCE), the first genuine scientist in history. Every scientist is in his debt.

The *Scientific Revolution* refers to the period in Europe between roughly 1550 and 1700, when modern science emerged. Developments in mathematics, physics, chemistry, astronomy and biology, including human anatomy, were the agents that brought historical changes in thought, belief, social and institutional organization.

The Scientific Revolution began with Nicholas Copernicus and ended with Isaac Newton. It influenced the *Enlightenment* or *Age of Reason* (1650s to 1780s), an era in which cultural and intellectual forces in Western Europe emphasized reason, analysis, and individualism rather than traditional lines of authority.

A few well known intellectuals from the Scientific Revolution:

Nicolaus Copernicus (1473-1543) was a Polish mathematician and astronomer who formulated the model of heliocentrism, that the Sun, rather than the Earth, is the center of the universe. Just before his death in 1543 Copernicus published De revolutionibus orbium coelestium (On the Revolutions of the Heavenly Spheres). It is considered to be a major event in the history of science.

Andreas Vesalius (1514-1564) was born in Brussels and is known for his work on human cadavers. He is considered the founder of modern anatomy and author of the seven-volume De humani corporis fabrica (On the Fabric of the Human Body), published in 1543.

Galileo Galilei (1564-1642) was an Italian physicist, mathematician, engineer, astronomer and philosopher. Galileo's championing of heliocentrism was controversial during his lifetime, a time when most subscribed to the idea that the earth was the center of the universe. In 1615 the *Supreme Sacred Congregation of the Roman and Universal Inquisition,* known as the *Roman Inquisition,* a system of tribunals developed by the Holy See of the Roman Catholic Church, concluded that heliocentrism was false and contrary to scripture. They placed works advocating the Copernican system on the index of banned books and forbade Galileo from advocating heliocentrism. Pope Urban VIII forced Galileo to recant and he spent the last nine years of his life under house arrest.

Thomas Hobbs (1588-1679) was an English philosopher. His 1651 book, Leviathan, established social contract theory, the foundation of most later Western political philosophy. Hobbes also developed some of the fundamentals of European liberal thought: the right of the individual, the natural equality of all men, and the artificial character of the political order. Hobbs would make the distinction between civil society and the state, the view that all legitimate political power must be "representative" and based on the consent of the people. This liberal interpretation of law would leave people free to do whatever the law does not explicitly forbid.

René Descartes (1596-1650) has been called the father of modern philosophy as well as the father of analytical geometry. His 1641 *Meditations on First Philosophy* continues to be a standard text at most university philosophy departments. He developed the *Cartesian Coordinate System,* allowing reference to a point in space as a set of numbers and allowing algebraic equations to be expressed as geometric shapes in a two-dimensional coordinate system and conversely, shapes to be described as equations. Most know him from his philosophical statement: "Cogito ergo sum." I think, therefore I am.

Antonie van Leeuwenhoek (1632-1723), a Dutch scientist, is known as the Father of Microbiology. He is best known for his work on the improvement of the microscope. Van Leeuwenhoek discovered and described red blood cells, blood flow in capillaries, spermatozoa, muscle fibers and an assortment of micro-organisms including bacteria.

Robert Hooke (1635-1703) was an English scientist, philosopher, architect and polymath. In 1665, using the microscope in exploration of microscopic life, he published the very popular book Micrographia, complete with engravings of what he observed. Hooke drew microscopic depictions of insects, a louse, a fly's eye and plant parts composed of what he called cells. Although the book is best known for demonstrating the power of the microscope, Micrographia also described the rotations of Mars and Jupiter, postulated the wave theory of light and the organic origin of fossils. His preliminary work was on gravity and the inverse square law, which governs the motion of the planets, further developed by Newton.

Issac Newton (1642-1727) was an English physicist. He is considered to be one of the most influential scientists of all time. His book Mathematical Principles of Natural Philosophy, 1687, laid the foundations for classical mechanics, formulating the laws of motion and universal gravitation.

The nineteenth and twentieth centuries had thousands of scientists producing hundreds of thousands of discoveries, all built upon the foundation of previous scientific research. It would be well beyond the scope of this modest appraisal of science to document all the achievements of these centuries. However, I must relate one story of Jonas Salk relating to his discovery of a vaccine for polio. Jonas Salk was a friend and mentor to me.

The development of a vaccine can only happen if scientists clearly understand the complexity of both the virus and the cellular components of the host, in most cases, humans. The development of Salk polio vaccine is an example of the effectiveness of scientific collaboration. Dr. Jonas Salk (1914-1995), the creator of the Salk polio vaccine, is known for his unselfish nature; he worked for the betterment of mankind without monetary incentive.

Until 1957, when the Salk vaccine was introduced, polio was considered one of the most frightening public health problems in the world. In order to test the effectiveness of the vaccine, field tests were set up involving 20,000 physicians and public health officers, 64,000 school personnel and 220,000 volunteers. Over 1,800,000 school children took part in the trial. When news of the vaccine's success was made public on April 12, 1955, Salk was hailed as a miracle worker. His sole focus had been to develop a safe and effective vaccine as rapidly as possible, with no interest in personal profit. When asked who owned the patent to it, Salk said, "There is no patent. Could you patent the sun?" In 1960, he founded The Salk Institute for Biological Studies in La Jolla, California. Today, The Salk Institute is a premiere center for biological research. Interestingly, one of the concerns of Jonas Salk was to create an institute that bridged the gap between science and the humanities. Salk said, "I first heard of Jacob Bronowski in 1960. My interest was aroused by recognizing the correspondence between his ideas and what I had in mind in trying to create an institution to heal the separation referred to by C.P. Snow as the 'Two Cultures'."

C.P. Snow (1905-1980) was a British physical chemist and novelist. C.P. Snow gave a 1959 Rede Lecture entitled *The Two Cultures.* Its thesis was that the intellectual life of the whole of western society was split into two cultures, namely the sciences and the humanities. and that this was a major hindrance to solving the world's problems. [29]

*Basic scientific research* is a systematic study directed toward greater knowledge or understanding of the fundamental aspects of nature. It is an adventure into the realm of the unknown. It arises out of the curious brain. The fundamental principle of basic research is to understand phenomena without monetary considerations, and the research may or may not have a specific human application.

*Applied scientific research* in contrast to basic research has a specific goal. Most often the research is profit-driven as it most likely has commercial applications. Modern medicine is built upon the work of basic science. Of the numerous new drugs available for the public, all are designed based upon the discoveries of the basic sciences.

Early man tried various herbs for medicinal purposes. Trial and error was the only method available to find which herb might be effective and which might be toxic. To understand which drugs may be effective, it is necessary to have knowledge of the components of a living organism, how each component functions and interacts with other components and with the external environment. The design of a new drug begins with a thorough understanding of not only all the life precesses but with a knowledge of chemistry and biochemistry. The sequence of DNA, the structure of the biochemicals, the properties of cellular membranes, the constituents of the metabolic cellular processes, the binding affinities of the various components to their intended target—these properties and more are required in the development of any drug. Drugs to be tested are not randomly selected. The process commences in the laboratory with basic research. A drug is obviously a chemical compound. Researchers may find that, within a specific chemical, merely moving a hydroxyl group from one carbon atom to an adjacent carbon may have dramatic effects. This small change may result in a compound that has a higher or lower affinity for the targeted site of action. Designing drugs is not haphazard. The design is calculated. New drugs are approved for usage only after rigorous government-approved testing and trials. Modern medicine is the result of basic scientific research.

*There is not any part of good government more*
*worthy than the further endowment of the world*
*with sound and fruitful knowledge.*

—Francis Bacon, Advancement of Learning, *1605* *ARISTOTLE*

The cost of new drugs developed by pharmaceutical companies is often recuperated by their subsequent sales. The costs of basic research has no such source of revenue. Money for basic research primarily comes from federal grants, usually the National Institutes of Health (NIH) and the National Science Foundation (NSF). The NIH budget request for 2016 is $31.3 billion. The NSF budget request for 2016 is $7.7 billion.

By comparison, total federal expenditures for 2016 are projected to be around $3.9 trillion. The US Defense budget is $622 billion, approximately 16% of the total federal government budget. NIH will receive 0.8% and NSF will receive about 0.2% of total federal expenditures for 2016.

Both the NIH and the NSF award grants to scientists who have submitted detailed grant proposals. These grants are reviewed, and only about 20% are funded. For the scientist, this means submitting at least five grant proposal to have a reasonable chance of receiving funding to continue research. [30]

Academic research is a major reason the United States remains a leader in medicine and biotechnology. Grants are the lifeblood of university research. Salaries must be paid. Research equipment is very costly. Until very recently funding for research had decreased 20% from 2004. The reason asserted by many is that the current US Congress has not much interest in spending more even on health research. There is, however, widespread bipartisan *public* support for governmental spending on basic research.

Younger scientists are especially hard-pressed to secure funding for their research. The rate for awarded first-time NSF applications fell from 22% in 2000 to 15% in 2006. In China their federal spending for basic research has increased threefold in the last decade. [31]

*Frontiers in Innovation, Research, Science and Technology* (FIRST) was a bill introduced in the House Science Committee May 28, 2014, by Republican Representatives Larry Bucshon and Lamar Smith. It would appear, by the title of this bill that it would be supportive of basic scientific research. Unfortunately the bill proposed a 40% cut in the NSF budget. These two Representatives argued that the NSF funds useless research that wastes taxpayers money. Fortunately for science and the nation, the bill was not enacted. The bill's supporters make the basic mistake of equating scientific research with spending. It's clearly not. Publicly funded research is particularly fruitful. Economists have found that every 0.5% spent on R&D returns 9.5% GDP growth. [32]

*We need to embark on a radical cultural shift in which science takes its rightful place alongside music, art, theater and literature as an absolutely indispensable part of a full life. We need to make clear that science is not something that you can willfully ignore. All of the major decisions going forward, from stem cells to nuclear proliferation to nanotechnology to genetically modified food to alternative energy sources to climate change, have a scientific component.*

—BRIAN GREENE (1963), AMERICAN THEORETICAL PHYSICIST AND STRING THEORIST. HIS BOOK THE ELEGANT UNIVERSE WAS A BEST-SELLER.

## Constructs: *The Humanities*

*If I could say it in words there would be no reason to paint.*

—EDWARD HOPPER (1882-1967) PROMINENT AMERICAN REALIST PAINTER AND PRINTMAKER.

The humanities are the noble expression of the human experience. Philosophy, literature, religion, art, music, history and language are understood to fall under the humanities umbrella. Knowledge of these human experiences gives us the opportunity to feel a sense of connection to those who have come before us as well as to our contemporaries. I intend to restrict this section of my treatise to the subject of art, that for which I have great passion and a greater acquaintance. The other disciplines of the humanities are no less important; I am unfortunately less familiar with them.

Humans do art because we are human. We are by nature creative. Art expression is a method of communication; we tell stories of our lives, our shared histories or experiences, both tragic and comical. Art chronicles our own lives and experiences over time. We need art to understand and share our history. The practice of art requires us to think, to be reflective and to be inspired. It stimulates our brains. Without art we do not have culture. Past cultures had art. We know past civilizations by their culture, by their artistic accomplishments, not so much by their politicians, kings, popes or rulers but by their artists. We know of Leonardo da Vinci, Michelangelo, Rembrandt, Claude Monet, Paul Cezanne and countless great artists from earlier times. Naming the Pope, King or Emperor of their times may be more difficult. We embody meanings into artworks. Art is the language that all people speak across racial, cultural, social, educational and economic barriers. Art enhances cultural appreciation and awareness. Art raises the ordinary to the extraordinary.

"We believe strongly that the arts aren't somehow an 'extra' part of our national life, but instead we feel that the arts are at the heart of our national life. It is through our music, our literature, our art, drama and dance that we tell the story of our past and we express our hopes for the future. Our artists challenge our assumptions in ways that many cannot and do not. They expand our understandings, and push us to view our world in new and very unexpected ways...That is the power of the arts—to remind us of what we each have to offer, and what we all have in common; to help us understand our history and imagine our future; to give us hope in the moments of struggle; and to bring us together when nothing else will. That is what we celebrate here today." —-Michelle Obama at Pittsburgh Creative & Performing Arts School, September 25, 2009.

The National Endowment for the Arts (NEA) was created by an act of the US Congress in 1965 as an independent agency of the federal government, *dedicated to supporting excellence in the arts, both new and established; bringing the arts to all Americans; and providing leadership in arts education.* Since its inception, the NEA has awarded more than 140,000 grants, including early support for the Vietnam Veterans Memorial design competition. For more than four decades the Arts Endowment has encouraged creativity through support of performances, exhibitions, festivals, artist residencies and other arts projects throughout the country. In partnership with the state and jurisdictional arts agencies and regional arts organizations, the NEA provides federal support for projects that benefit local communities. A very few examples include support for the Sundance Film Festival and the Spoleto Festival USA held in Charleston, South Carolina, which features performances by renowned artists as well as emerging performers in opera, theater, dance and chamber, symphonic, choral and jazz music. Additional support has been provided to the Young Authors Book Program, an in-school literary arts program for underserved high school students. Other grants have been given to the PBS's Great Performances series and the American Film Institute.

Since its conception there have been numerous conservative attacks aiming to abolish the NEA. Ronald Reagan attempted to abolish the NEA but was persuaded not to do such by his conservative friends in Hollywood. A period in the late 80s and 90s witnessed abolishment attempts by Pat Robertson, Dick Armey, Pat Buchanan and Donald Wildmon of the American Family Association. Newt Gingrich attempted to eliminate not only the NEA but also the National Endowment for the Humanities and the Corporation for Public Broadcasting. The artists that most offended these people included Andres Serrano, Karen Finley and Robert Mapplethorpe. In 1990 the act governing the US National Endowment for the Arts was amended, requiring that the judges of grant applications should take into consideration *general standards of decency and respect for the diverse beliefs and values of the America public.* The US Supreme Court found the provision to be constitutional.

*Entartete Kunst* was the title of an exhibition produced by the Nazi regime in Munich, Germany, in 1937. The English translation is *Degenerate Art,* a term used by the Nazis to describe all modern art that was banned on the grounds that it was un-German or Jewish Bolshevist in nature. The paintings in the *Degenerate Art* exhibition were modernist artworks displayed in a chaotic manner accompanied by disparaging labels. The purpose of their exhibition was to inflame public opinion.

These *degenerate* artworks were in opposition to the regime's insistence that art should exalt Nazi ideals: racial purity, militarism and obedience. During the Nazi period over 5,000 artworks were seized, including works by Nolde, Heckel, Kirchner, Beckmann, Archipenko, Chagall, Ensor, Matisse, Picasso and van Gogh. Artists identified as degenerate by the Nazis were subjected to sanctions: they were dismissed from teaching positions and forbidden to produce, exhibit or sell their art.

The August 20, 2013, issue of the *New York Times* reported that the House Appropriations Committee has cut the budget of both the National Endowment of the Arts and the National Endowment for the Humanities by one-half. Over the last 40 years, the NEA has survived a dozen conservative onslaughts in Congress attempting to dissolve it completely. [33]

The 2013 budget of $146 million for the NEA represents just 0.004% of total federal expenditures. The National Endowment for the Humanities is similarly funded. The NEA has sustained significant budget reductions. NEA appropriations have declined by $21.5 million (-13%) in the last three federal budgets. Each $1 in NEA grant funds leverages an additional $9 from other public and private sources. The arts create jobs. The nonprofit arts and culture industry creates an economic impact of $135 billion and supports 4.13 million jobs in multiple industries. The arts generate revenues. The arts sector generates $22.3 billion in federal, state and local tax receipts. [34]

Nick Gillespie, a libertarian American journalist, in the *Economist*, August 29, 2012, argued that governmental support of specific institutions or individuals is in no way necessary or sufficient for the production of "art" (his quotation marks). His second objection was that "governments everywhere are dead broke. Further: Forced funding of the arts—in whatever trivial amounts and indirect ways—implicates citizens in culture they might openly despise or blissfully ignore. And such mandatory tithing effectively turns creators and institutions lucky enough to win momentary favor from bureaucrats into either well-trained dogs or witting instruments of the powerful and well-connected.

Independence works quite well for churches and the press. It works even more wonderfully in the arts."

So Nick Gillespie, would you consider *The Vietnam Veterans Memorial*, designed by American architect Maya Lin, a waste of federal money? So simple, so elegant, so expressive.

America's success was built on a liberal arts education. Liberal arts subjects, such as creativity, aesthetic sensibility, English, philosophy and political science teach people how to think, write and communicate. It's important to try to educate people in basic concepts no matter what their eventual specialization.

—Fareed Zakaria (b. 1964), Indian-born American journalist and author. He is the host of CNN's "Fareed Zakaria GPS" and writes a weekly column for *The Washington Post.* [35]

It seems that some members of Congress fear the marketplace of ideas, artistic expression and creative innovation precisely because they can't control them. At worst, that's a form of fascistic thinking, and reason enough in my mind to support the NEA. At very best, it's using personal values to assess the arts, by which some members of Congress confuse personal morality with public policy. Even more, the NEA funnels 50% of its money directly to state arts agencies (by congressional mandate), thereby allowing local standards and selection criteria to prevail (I'm not sure everyone in Congress understands this part). Also, the NEA primes the pump for private giving to the arts by individuals, foundations and corporations in amounts which dwarf the NEA's own budget by thousands of times.

—Jonathan Abarbanel, Dept. of Theater, University of Illinois at Chicago. August 16, 2011.

Funding public artworks is a very contentious topic. Artists, humanists and Liberals believe that there is a need for public support of the arts. They are an expression of our civilization, a historical statement of the noble expression of our humanity.

Conservatives see no need for public support for art, believing instead that the arts should be self- supporting, often with the accompanying remark: *I don't like it, I don't understand it and I don't want to pay for it.*

Many municipalities are now requiring developers to set aside a percentage of construction costs, usually about 1%, for public art.

Arguments favoring governmental support of the Arts cover several themes. Often sighted is that Art Education, exposing youngsters to art, fosters imagination and facilitates the child's educational success. It is a way for disadvantaged children to experience education in a positive and creative manner. Art creates jobs and produces tax revenues, especially from tourists. Public art creates a sense of place and history and speaks to the quality of the human endeavor. With two million full-time artists and nearly six million art-related jobs in this country, arts jobs are real jobs that are part of the real economy. Art workers pay taxes and art contributes to economic growth, neighborhood revitalization and the livability of American towns and cities. Public art invigorates public spaces. It can make us pay attention to our civic environment. It leaves a legacy for the future.

*The strongest defense of the humanities lies not in the appeal to their utility—that literature majors may find good jobs, that theaters may economically revitalize neighborhoods—but rather in the appeal to their defiantly non-utilitarian character, so that individuals can know more than how things work, and develop their powers of discernment and judgment, their competence in matters of truth and goodness and beauty, to equip themselves adequately for the choices and the crucibles of private and public life.*

—Leon Wieseltier (1952), American writer, critic, and literary editor (1983 - 2014) of *The New Republic.*

*Man's inhumanity to God. The pain we cause him. We've poisoned his atmosphere. We've slaughtered his creatures of the wild. We've polluted his rivers. We've even taken God's noblest creature, Man, and brainwashed him into becoming our own product, not Gods, hacked, stacked and canned.*

Spoken by Richard Burton as the character Rev. Shannon from *The Night of the Iguana,* 1964.

# Seven

## Parrhesiastes' Assessment on The Nature of Humans

The Nature of Humans is a complexity of genetic traits, behaviors and human constructs. There cannot be an assessment that all would concur. There will always be a bias. With the edict of Dr. Garrett Hardin: "Be critically observant. Be rational and intellectually honest" and the judgment of Parrhesiastes, here is an assessment on the nature of humans. Hopefully this will lead to a greater level of enlightenment and a positive future for humanity.

*The creation and evolution of life on the planet Earth has been extremely successful against all odds. Not really. Life is the inevitable occurrence when specific conditions occur. The Big Silent Bang and the expansion of the universe produced the chemical elements: carbon, hydrogen, oxygen, sulfur, the halogens, iron, magnesium and all the other elements found on the Periodic Table. Carbon is the basic building block of life. Carbon has 4 out of 8 possible electrons in its outer shell, perfect for accepting four more electrons from other elements, oxygen, hydrogen or itself. An electrical discharge and presto, life, an extensive assemblage of self-reproducing biochemicals.*

*Earth has never been a static body in space. It has had very turbulent early years: shifting tectonic plates, volcanic eruption, and earlier cosmic collisions with asteroids and cosmic debris. Life originated on the planet Earth 3.6 billion years ago under conditions that presented geological challenges to life forms. In four to five billion years the sun will swell up and become a red giant as big as the Earth's orbit. All life on earth will be incinerated. When Homo sapiens will cease to exist is certain but unknowable. The intervening years will be very interesting. Perhaps these animals are approaching their term limits.*

*Millions of species of animals and plants have evolved on Planet Earth. The diversity of life forms is truly amazing. Plants are quite remarkable. Plants are solar-powered thanks to their chlorophyll. They are essential to H. sapiens for numerous reasons. Plants supply them with food. Plants supplied all the oxygen requirements for H. sapiens as well as all other oxygen- consuming animals. Plants have supplied almost all their energy, wood and fossil fuels. Plants also remove carbon dioxide from the atmosphere but not as rapidly as H. sapiens are putting it back. Plants simply can't keep up. The oceans are helping to remove atmospheric carbon dioxide but the oceans are turning acidic in the process. In a fashion similar to H. sapiens, plants partake in chemical warfare. Plants evolved toxins to ward off predators. Some plant chemicals have become phyto-pharmaceuticals. H. sapiens really appreciate some of these drugs. Plants have the ability to reproduce offspring in great numbers, as H. sapiens have. Plants have developed clever techniques to disperse their offspring, often by wind power. H. sapiens have likewise dispersed, most powered by plant-derived fuels. Plants have been able to prosper quite well on this planet. They seem to all get along. Fortunately for plants, they never developed the high level of cognition of H. sapiens. Therefore there is very little possibility of self-destruction. Bravo to plants.*

*Plants and animals share the same DNA, even some genes. Both have evolved by known evolutionary principles. The lineages are well established. The fossil records are comprehensive. There is no doubt. The Evolutionary Tree is virtually complete; there are just a few rather minor details still missing.*

*H. sapiens have certainly evolved over the last six million years. They still are animals and not at times too dissimilar from the pejorative meaning of being an animal. The brain developed complexity over a time scale measured in billions of years. The simple roundworm has 302 neurons. H. sapiens have 86 billion neurons. With mutation, selection and a bit of luck the marvelous H. sapiens brain evolved. Over the last six million years the brain has tripled in volume, especially the frontal lobe associated with memory, planning, reasoning, creativity and problem-solving. This expansion led to their high level of cognition. With all that complexity developed over millions of years, one would think the brain would be perfect. A perfect brain would be able to see and hear the world exactly as it is. Memory would not be fugitive. Judgments would be contemplative.*

*The claustrum is at the center of consciousness. The claustrum is involved in multi-sensory integration and synchronization of the two hemispheres of the cerebral cortex. The consequence of this activity is a seamless conscious experience. The claustrum is that organ that should enable H. sapiens to foresee the future.*

*There is one small brain structure, the amygdala, at the center of their emotional response. For some individuals the amygdala functions at a level consistent with compassion and kindness while for others the amygdala elicits behavior that can be raucous and belligerent. Without the amygdala they would not have a fear response in the face of danger.*

*The brain of H. sapiens is not a blank slate at birth. The brain comes already programmed with traits that originated with their ancient ancestors. H. sapiens all appear similar, yet there are some trivial physical differences such as skin and hair color; otherwise they are mostly homogeneous. The exception to this homogeneousness is the variation in the level of expression of their genetic traits. Immediately reacting with fear, greed or prejudice is a primitive survival trait. These traits are often the antecedent to war. Their reasoning is often motivated, not always logic-based. Fear is extremely effective when used by politicians. These creatures have an adage: Create fear, get elected.*

*Morality, empathy and compassion are also genetic traits that lead to peaceful living. Rational thinking with intellectual honesty distinguishes enlightened individuals from the unenlightened. Contemplative consideration requires considerable effort, which is beyond the ability of most H. sapiens.*

*What they seem to have ignored is that these animals are governed not only by their by their experiences, teachings and social norms, but also by their ancient survival genes. All thoughts, decisions, schemes, shenanigans and creativity are the result of brain activity. Often unappreciated is that genetic variants of their biological traits contribute to some bizarre behaviors. Behaviors displayed are dependent upon which isoform of GABA, DARPP-32 or an assortment of other biochemicals are inherited. Serotonin or dopamine concentrations and their regulatory constituents may vary between individuals. Some bizarre behaviors are a combination of several genetic traits. One is designated as the Cheney-esque Behavior, which combines fear, anger and greed into one despicable behavior. Another behavior combines delusion and motivated reasoning designated as the Inhofe-esque Behavior. These two behaviors are often displayed by politicians.*

*While there is no universally acceptable definition of Free Will, most H. sapiens believe they do have Free Will. It is debatable whether H. sapiens can act freely. The decisions they make are dependent on the interface of structures, biochemicals and neural networks that comprise their brain as well as their environment, prior knowledge, and social constraints.*

*Most H. sapiens are honest and ethical, most want to be productive, most are creative and many have concerns for the rest of humanity. Creativity, the sciences and the humanities are their most noble traits. Not all creativity has been beneficial for humanity's longevity. With their advanced creativity their spears have greatly improved. Now a single spear can travel thousand of miles and kill tens of thousands.*

*These animals consider that they are civilized. They have created great empires, great architecture, great literature and art. There have been numerous civilizations. Many were built by the work of low paid workers or by slave workers. Most all had large armies. Considering their propensity for killing each other, denigrating others that are dissimilar, one would question what they mean when they say that they are civilized. That is the question that they should ask themselves. Are these animals capable of building a civilization where equality has deep meaning?*

*No other lifeforms have the ability to harm others of their own kind and to alter the environment as have H. sapiens. In this they are unique. H. sapiens have a proclivity to alter the very planet they rely upon for survival. Their planet is alarmingly warming. They have failed to realize the importance of a balanced and sustainable ecosystem. Most H. sapiens seem not to have the ability to acknowledge and to intellectually accept that there are serious environmental impediments in their future, a consequence of their nature.*

*There seems to be a compulsion among many to assume that life cannot exist without a creator and a purpose. To this end these H. sapiens invented a higher authority. This higher authority arose out of the trait of self-realization. With self-realization their inquisitive nature increased. An often asked question is: What is the purpose of life? For answers, early H. sapiens constructed religions within their own tribe. Wealth accumulation is a sufficient purpose for many. Others strive for accomplishments that are more widely appreciated.*

*Within some tribes, member may speak directly to this higher authority but only the few who are ordained have a true two-way conversation. The ordained convey the words of this authority to their disciples, the obligation of obedience. This is a brilliant concept. Power is kept within the elite and the tribes remain united. This framework allows the ordained to justify as needed the tampering of the biosphere and the termination of non-tribal members.*

*All life may be considered special; however, H. sapiens believe that they are extra special. They consider themselves to be a noble species with a bright future. They are not that special. This species is similar to all the other species inhabiting earth: same DNA and same basic biochemistry. They believe that they can solve every problem. This is not likely. Not because they lack the intellectual prowess; rather it is because they can be an unruly tribal species and their behavior is linked to their inherited genetic traits. They seem not to be able to just get along with each other. Finding commonality between their tribes has proven to be beyond their capability. Shouting, not logical and reasoned debate, seems to be their preferred of method intertribal communication. Between disparate tribes war is their ultimate method of communication and this method is widely used. Tribal loyalty is so pervasive that compromise is difficult.*

*H. sapiens can be contrarians. Copernicus and Galileo were contrarians. Duchamp and Picasso were contrarians, even Monet. These contrarians went against conventionality. Their achievements were the result of questioning and research. Their achievements have been time- tested. They were unique. Only the well prepared are able to make these discoveries and innovations.*

*Within their species there are a multitude of contrarians much different from these pioneers. Most of these individuals skip the research phase and proceed directly to their revelations. H. sapiens sent men to the moon and returned them safely. There are deniers who contest this achievement with the assertion that it was staged in a Hollywood studio. Creationists rely upon ancient scripture as factual. The Kennedy assassination and 9/11 brought conspiracy advocates out in groves, with theories that the events didn't happen as widely reported. There are climate change deniers with conspiracy theories all their own. Many are contrarians simply because of tribe loyalty, not as a result of reasoned investigation.*

*This contrarian behavior is one reason democracy is in such disarray. There are too many contrary opinions as to the better course of governance. And there are too many people with an excessive greedy constitution that place their own needs ahead of the rest of humanity. These are impediments to democracy. Opinions are often rendered with self-interest. Few choices are made with contemplation and consideration of humanity's long-term requirements. Democracy is intended to give everyone's opinion equal weight. Presently this is not happening. In theory democracy should self-correct. The continuing disparity of wealth may tip the balance past the point where an equilibrium can be achieved. Change may happen; the rich may become compassionate. Most likely there will be uprisings and the subsequent degradation of democracy.*

*H. sapiens are a tribal species. They are very defensive of their own tribe. They have the ability to discern differing characteristics of a foreign tribe. This ability often contributes to conflict. Zealot supporters often have a propensity for war and killing. They abhor bloodshed of their own tribe members yet seem to relish bloodshed of outsiders. Peace and harmony, while earnestly desired by most all tribes, is unobtainable. Most conspicuous in this species is that there are two tribes identifiable by their attitudes towards moral codes, social and economic values. These two major tribes are the Conservative tribe and the Liberal tribe.*

*When comparing and contrasting these two tribes, there are some general characterizations. These will not be absolutes, but generalities that may be helpful in identifying the values and mindsets of these tribes. They can be summarized as:*

*Conservatives are often cautious and hold to traditional values. They can be sober and conventional, avoiding novelty or showiness. They are disposed to preserve existing conditions and institutions. Conservatives believe that censorship is necessary to ensure that moral standards are met. Their nature is exemplified by their resistance to civil rights and other social engineering programs.*

*Whereas—*

*Liberals are open to new behaviors and opinions and have a willingness to discard traditional values. They are favorable to progress or reform, especially the freedom of the individual and governmental guarantees of individual rights and liberties for all. Generally they are free from prejudice or bigotry, especially with respect to matters of personal belief or expression.*

*For an orderly society, Conservatives often believe that moral guidance should come from a higher source, from ancient scriptures.*

*Whereas—*

*Liberals insist that behavior should be governed by reasonable manmade laws that contain safeguards to protect all citizens and all life forms.*

*Conservatives believe that corporations are the life blood of a country. They should be self-regulated. They create jobs that build the weapons to defend the country. Capitalism and self-interests are essential for prosperity. Of course corporations should have a vote. Regulations should be on individuals rather than businesses. Gerrymandering and voter suppression is necessary for maintaining a majority.*

*Whereas—*

*Liberals want restrictions on corporations to prevent fraud and the manipulation of democracy. They believe that corporations are like viruses; they infect governments. They do not have DNA, but something better, money. Democracy as intentioned is not functioning well, held captive by the elite, an oligarchy, not the electorate.*

*Conservatives believe we should build more prisons and maintain long prison sentences for felons. This pumps the economy, keeps the public safe and insures that many blacks and latinos aren't allowed to vote. Those groups primarily vote for Liberals.*

*Whereas—*

*Liberals believe that the judicial system needs reform, with reasonable sentencing and a stronger focus on rehabilitating prisoners.*

*Wealth is the measure of success for Conservatives. Higher education should be paid for by the recipient of education. Humanity's purpose is to exalt the highest ideals of the Conservative tribe.*

*Whereas—*

*A high level of education is the measure of success for a liberal and it should include the sciences and the humanities. Liberals believe that society should provide free higher education as it provides long-term economic gains. Educated people are needed to rebuild the failing infrastructure thus creating jobs and consequently reduces crime. Science and the humanities are an expression of their creativity.*

*Conservatives believe that taxes should be low or non-existent, Free Coke so to speak. Taxes should primarily be used to strengthen our military-industrial establishment and not much more.*

*Whereas—*

*Liberals believe that taxation is the money citizens are assessed for governmental services that benefit all citizens.*

*For Conservatives the environment is a source of unlimited resources. Conservatives believe that if the earth is warming, it would present significant economic opportunities such as the opening of the arctic for shipping and oil extraction.*

*Whereas—*

*Liberals believe the environment should be protected for future generations. Liberals are concerned that anthropogenic climate change is leading to dire consequences.*

*There are commonalities. The more strongly one is affiliated with a tribe, the more likely their reasoning becomes motivated. Strong adherents are likely to ignore facts, more likely to preach and recruit, and more likely to condemn outsiders. Both tribes want what's best. Both tribes want to change the other tribe's behavior. Best can only be assayed against a standard. There is no universally recognized metrological behavioral standard for best behavior.*

*Neither tribe has exclusivity on compassion and empathy or anger and greed. These and other traits reside within both tribes to varying degrees. Some of these traits are more prevalent in one tribe than the other tribe. That which is without doubt is that the tribes would not agree as to which traits are prevalent within their own tribe. It cannot be declared which tribe's values and behaviors are better.*

*Only time will determine which values and behaviors will endure, which tribe's values will lead to long-term harmonious survival.*

*Another commonality among most all the tribes members is the belief that there is a purpose to life. For some it is the simple purpose of providing essentials for their offspring. This trait surely has a genetic component. For many other individuals their purposes is to accumulate great wealth. Others aspire for what they consider lofty humanitarian ideals. There is as strong cultural tradition for such purposes.*

*The Greeks were the founders of the rational and humanistic tradition. For these ancient Greeks intellectual thought does not come from myth-based religion. Nature is not manipulated by arbitrary and willful gods or by blind chance. The Age of Enlightenment challenged faith, tradition, superstition and intolerance and abuses of power by the church and state. Some current faiths and political parties would prefer for humanity to return to the Dark Age's sensibility.*

*H. sapiens' solar system may not be the only planetary system with life. There may have been other attempts at life. There are certainly other planets in the universe that are Goldilocks planets, in an appropriate position within a solar system, not too far or too close to their star. Some planets may have the proper constituents for life formation: the proper chemicals, a slight tilt for seasonal changes, and a competitive system that selects the fittest. There may have been other universes prior to their Big Bang. There could be future universes after the collapse of this universe. Would any of these presumptive universes do better? Future worlds would need to start with the same building blocks, the ones found on the Periodic Table. All life would be subject to the same physical restraints. The H. sapiens species rose to such dominance as to have the ability to destroy themselves. Would "new" life forms be tolerant and not participate in slavery, wars or mass shootings? Would the next Universe be more successful? Maybe this is already the fourth or fifth universe in an endless string of universes. Most likely any past or future system would be similar to the present universe. One might speculate that H. sapiens have been participants in a grand biology experiment.*

*This species calls itself humans. That they are humane is debatable. Undeniably, the nature of humans is fascinating.*

*This assessment includes my counsel:*

*Homo sapiens, you are on your own. There will be no outside help. There is no higher authority. Act smartly, logically and with consideration of all consequences. Arrogance clouds judgment. Fear impedes reasoning. Turn fear into concern. Be wary of bombastic nincompoops. Adopt the sensibilities of the Greeks and the scholars of the Age of Enlightenment. Honor scholastic achievements. Honor the humanities. Be a humanitarian. Respect your environment. The planet is warming. Do the obvious. The disparity of wealth threatens democracy. Strive for equality for all. Shame the despicable behaviors of war and punishment. Endeavor to be intellectually honest. Emulate the Bonobos! Change to a matriarchal society. Reduce aggression, resolve conflict and stress, with a zeal for constant high level of sexual behavior. If you do, be sure to use contraceptives; the earth is already over populated.*

*And that's the way it is*

# Epilogue

The earth has been constantly changing. Tectonic plates rearranging land masses, cosmic impacts, the ebb and flow of ice sheets and volcanism have drastically altered this planet. Over the last ten million years the landscape of East Africa changed from flat homogenous areas covered with tropical forest to heterogeneous regions with grassy tree-dotted savannas, deserts and mountain regions of cloud forests. These regions prevailed at most East African sites where hominin speciation commenced six million years go. Speciation occurred when tribes exceeded an optimal size, and splinter groups separated into new tribes and spread throughout Africa and eventually the rest of the habitable planet. Within this diverse and challenging environment these early tribal hominins evolved, either forming new species with the necessary survival traits and multiplied, or they perished. Only one hominin species remains. Modern *Homo sapiens* remain tribal animals, and it should not be surprising that we share the behavioral genetic traits of our ancient ancestors.

The planet is warming and we are a contentious species. As the planet warms there will be major disruptions to human life. Tens of millions will be displaced by worldwide coastal flooding. There will be human suffering due to shortages of food and water. It is the nature of humans to do what is needed to survive. It is the nature of humans to start wars. Inevitably global warming will lead to global conflict. It is also in our nature to be innovators, builders and humanists.

Our behaviors are due in part to our ancient genetic traits. Greed and tribal loyalty often take precedence over our humanitarian nature. We believe that we are able to arrive at the truth through our ability to be rational, yet we know that what we perceive is an interpretation of reality, filtered by our brain. Our reasoning is too often motivated by fear and tribal loyalty. Excessive fear clouds judgment. To decrease the severity of the impending catastrophe there needs to be a collective strategy devoid of religious or political ideology. Our political and religious views are deeply entrenched. With two prominent tribes with dissimilar sensibilities, can human-induced catastrophes ever be averted? Do we truly have free will? Is "peace and harmony" only a poetic phrase?

My aspiration for this book is to stimulate thought and discussion by highlighting our long history, human traits and behaviors. There is great diversity among our species, but we should seek the common good. Your values may differ from mine. I am reasonably certain I will not have changed many behaviors or beliefs. Whereas hope may only be a placebo, there are avenues that rational and creative individuals can pursue. Compassion, science and our creative traits can help humanity towards a path that will ensure long-term harmonious existence on Planet Earth. We need not become victims of our inherited behavioral traits. The future of mankind is unknowable, but there are indicators that we should pay attention. Will *Homo sapiens* thrive or wither? Consider the counsel of Parrhesiastes. Consider what your purpose is within the relatively short time of your life. How are you contributing?

# Acknowledgments

The genesis of this book commenced upon the completion of my book, *Homo sapiens* - A Liberals Perspective, 2014. This book fell short of what I had aspired. It needed to be revised, expanded and extensively re-written. This present book is the revision.

As with most people, my perception of the world unfolded slowly. I must have been about three or four years of age when one dark evening my father pointed up at the shy and said: Look, there is the Big Dipper and the Milky Way. It required explanations as to what he meant. I just saw stars. Most humans want explanations and too often partial explanations are superficial. I began to realize at an early age that most phenomena have complexity. This early awareness was life-changing.

My science classes provided avenues for further explorations. There is always more to learn and simple explanations, while offering some degree of satisfaction, are often inadequate, sometimes even wrong. I wondered why humans behave as we do. Why are humans so contentious? We all have the same DNA yet we behave so differently. As we learn more about the brain and behavior what is certain is that explanations are very complex.

I would like to make tributes to some of those individuals that have contributed to my understanding, awareness and knowledge. I am grateful for the discipline of Mr. Williams, my Junior High School science teacher and Mr. Ball, my High School botany teacher. At UCSB I was fortunate to have science classes taught by Garrett Hardin, Bob Haller, Wally Muller and Eduardo Orias. UC Santa Barbara is a Liberal Arts university and as such I took classes in art, music, philosophy english literature, and history with professors who names are now forgotten but their contributions remain solid.

I started my science career at The Salk Institute at the time when The Institute was founded. It was smaller and more intimate. Common were stimulating lunch time conversations with my science colleagues. Notable participants in these informal gatherings were Jacob Bronowski, Francis Crick and Jonas Salk. These individuals were the giants of science. Collectively they were the inspiration for my books. Of importance to Jonas Salk was that science should have a connection with the humanities. To that end he invited Jacob Bronowski to be a founding member of The Institute.

After my career at The Salk Institute I founded and curated The Bronowski Art&Science Forum as a tribute Jacob Bronowski and to revitalize Jonas Salk's idea to connect art and science. From 1999 through 2012 The Forum hosted 120 presentations, crucibles of conversations, by renowned scientists and artists. The Forum attendance grew to over 350 attendees. There were many individuals who were helpful in the success of The Forum. Patricia Frischer aided in publicity, Kas Maslanka provided technical support, Steve Link provided many contacts with prominent scientists and the late Rita Bronowski, a consistent supporter, attended over 100 of The Forums.

I would like to acknowledge the contributions of individuals who have read the manuscript at various stages and made helpful contributions: Roger Guillemin and Dave Schubert of The Salk Institute, John Chalmers, Scripps Institution of Oceanography, UCSD Professor Stuart Anstis, Jacques Perrault, SDSU, Mark Stabb and artist Jason Rogalski. Special thanks of to Claire Slattery for proofreading the manuscript.

Many others have commented on this book in various stages of completion and offered valuable suggestions. These would including Robert Poe, Robert Anderson, Murray Powers, Jeanne Cretois-Burk and Dolores Welty. I especially want to thank my 20 year companion, Beverly Boggs, for encouraging me and for tolerating the many long day and endless weekends that I spent writing. Quotations are from a variety of sources including authors' words as published, brainy quotes.com, goodreads.com/quotes and searchquotess.com.

I am thankful for the enormous contribution that Wikipedia provides. It survives by contributions from uses.

# References


## 1
## Introduction

1. Foucault, M. *Parrhesiastes.* 1983, Six Lectures by at University of California, Berkeley.

## 2
## History

1. *Wikipedia,* Miller–Urey experiment

2. Douglas Fox, "Primordial Soup's On: Scientists Repeat Evolution's Most Famous Experiment" *Scientific American.*( March 28, 2007)

3. *Wikipedia,* Panspermia

4. Panspermia-Theory.com

5. Lee Sweetlove. "Number of species on Earth tagged at 8.7 million" *Nature-online 23* (August 2011)

6. Yeast Book, *Genetics,* November, 2011

7. *Wikipedia*, is an excellent first step as well as the websites of the Smithsonian, Australian Museum and TalkOrigins

8. "Bonobo & Congo Biodiversity Initiative". *BonoboConservation.com*

9. Kate Wong "Tiny Genetic Differences between Humans and Other Primates Pervade the Genome.". *Scientific American* Volume 311, Issue 3.

10. Prüfer,K., Munch, K., Hellmann,I., Akagi,K., Miller, Walenz, B., Koren,S., Sutton,G., Kodira,C., Winer,R., Knight,J., Mullikin, Stephen, Meader,J., Ponting,C., Lunter,G., Higashino,S., Hobolth,A., Dutheil,J., Karakoç,E., Alkan,C., Sajjadian,S., Catacchio,C.R., Ventura,M., Marques-Bonet,T. and Eichler, E. "The bonobo genome compared with the chimpanzee and human genomes." *Nature* 486, 527–531 (28 June 2012).

11. Michael Balter, M. "World's oldest stone tools discovered in Kenya.", April 14, 2015 *Science*

12. Harmand,S., Lewis, J., Feibel, C., Lepre, C., Prat, S., Lenoble, A., Boe,X., Quinn, R., Brenet, M., Arroyo, A., Taylor, N., Clement,S., Daver, G., Brugal, J-P., Leakey, L., Mortlock,R., Wright, J., Lokorodi, S., Kirwa, C., Kent, D., and Roche, H., "3.3-million-year-old stone tools from Lomekwi, West Turkana, Kenya." *Nature* vol 521 (May 21, 2015) pp. 310 - 326.

13. Dennis, M., Nuttle, X., Sudmant, P., Antonacci,F., Graves, T., Nefedov, M., Rosenfeld, J., Sajjadian,S., Malig, M., Kotkiewicz, H., Curry, C., Shafer, S., Shaffer, L., de Jong, P., Wilson,R., and Eichler, E. "Evolution of Human-Specific Neural SRGAP2 Genes by Incomplete Segmental Duplication.", *Cell,* Volume 149, Issue 4, p 912–922 11 May 2012.

14. Villmoare, B., Kimbel, W., Seyoum, C., Campisano, C., DiMaggio, E., Rowan, J., Braun, D., Arrowsmith, JR., and Reed, K., "Early Homo at 2.8 Ma from Ledi-Geraru, Afar, Ethiopia." *Science,* March 4, 2015

15. Paul HGM Dirks, Lee R Berger, Eric M Roberts, Jan D Kramers, John Hawks, Patrick S Randolph-Quinney, Marina Elliott, Charles M Musiba, Steven E Churchill, Darryl J de Ruiter, Peter Schmid, Lucinda R Backwell, Georgy A Belyanin, Pedro Boshoff, K Lindsay Hunter, Elen M Feuerriegel, Alia Gurtov, James du G Harrison, Rick Hunter, Ashley Kruger, Hannah Morris, Tebogo V Makhubela, Becca Peixotto, Steven Tucker. "Geological and taphonomic context for the new hominin species Homo naledi from the Dinaledi Chamber, South Africa." *eLife* 2015;4:e09561, September 10, 2015

16. Bradford Foundation, *BradshawFoundation.com/origins*

17. Ambrpse, S. "Late Pleistocene human population bottlenecks, volcanic winter, and differentiation of modern humans." *Journal of Human Evolution*, [1998] 34, 623-65

18. Meyer, M., Fu, Q., Aximu-Petri, A., Glocke, I., Nickel, B., Arsuaga, J., Martínez, I., Gracia, A., Bermúdez de Castro, J., Eudald Carbonell, E., and Pääbo, S., "A mitochondrial genome sequence of a hominin from Sima de los Huesos." *Nature* 505, 403–406, 16 January 2014.

19. Bryner, J. "Some Neanderthals Were Redheads.." *Live Science,* October 25, 2007.

20. Pearce, E., Stringer, E., Dunbar, R., "New insights into differences in brain organization between Neanderthals and anatomically modern humans." *Proceedings of the Royal Society B.* 13 March 2013.

21. Dediu, D. and Levinson, S. "On the antiquity of language: the reinterpretation of Neanderthal linguistic capacities and its consequences.." *Frontiers in Language Sciences* 05 July 2013.

22. Sankararaman, S., Patterson, N., Li, H., Pääbo, S., and Reich, D., "The Date of Interbreeding between Neandertals and Modern Humans.", *PLOS Genetics,* October 4, 2012.

23. Vernot,B., Serena Tuccil,S.,,Kelso,J., Schraiber,J.,Wolf,A., Gittelman, R.,, Dannemann,M., Grote,S., McCoy,R., Norton,N., Scheinfeldt,L., Merriwether,D., Koki,G., Friedlaender,J.,Wakefield,J., Svantc Pääbo,S., Akey1,J. "Excavating Neandertal and Denisovan DNA from the genomes of Melanesian individuals" *Science,* 17 Mar 2016

24. Cooper, A. and Stringer, C. "Did the Denisovans Cross Wallace's Line?" *Science* 18 October 2013: Vol. 342 no. 6156 pp. 321-323

25. Rendua, W., Beauvalc,C., Crevecoeurd, I., Bayled, P., Balzeaue, A., Thierry Bismuth, T., Bourguignong, L., Delfourd, G., Faivred, J., Lacrampe-Cuyaubèrec, F., Tavorminac, C., Todiscoj, D., Turqd,A. and Maureilled, B. "Evidence supporting an intentional Neandertal burial at La Chapelle-aux-Saints" December 12, 2013. *Proc Natl Acad Sci USA,* vol. 111 no. 1, 81–86.

26. Hall, S. "Last of the Neanderthals" *National Geographic. October, 2008*

27. Higham, T., Douka, K., Wood, R., Bronk Ramsey, C., Brock, F., Basell, L., Camps, M., Arrizabalaga, A., Baena, J., Barroso-Ruíz, C., Bergman, C., Boitard, C., Boscato, P., Caparrós, M., Conard, N., Draily, C., Froment, A., Galván, B., Gambassini, P., Garcia-Moreno, A., Grimaldi, S., Haesaerts, P., Holt, B., Iriarte-Chiapusso, M., and Jelinek, A. "The timing and spatiotemporal patterning of Neanderthal disappearance." 21 *Nature* 512, 306–309. August 2014

28. Hardin, G. "The Competitive Exclusion Principle" *Science,* Vol. 131, No. 3409 Apr. 29, 1960, pp. 1292-1297.

3

Brain

1. Gorman, J. *New York Times, Science,* November 10, 2012

2. Gorman, J. "The Brain's Inner Language." *New York Times, Science,* Feb 24, 2014

3. Crick, F. The Astonishing Hypothesis p. 3, Charles Scribner's Sons, 1994.

4. Randerson, J. *The Guardian,* 28 February 2012

5. Shuo, W., Oana, T., Mamelak, A., Ross, I., Adolphs, R., and Rutishauser, U. "Neurons in the human amygdala selective for perceived emotion." *Proc Natl Acad Sci USA,* 111 (30) 2014.

6. Deacon, T. "A role for relaxed selection in the evolution of the language capacity." *Proc Natl Acad Sci USA,* 9000-9006, 2010.

7. Klein, R. and Edgar, B. The Dawn of Human Culture: a Bold New Theory on What Sparked the "Big Bang" of Human Consciousness. John Wiley & Sons. 2002

8. Cherry, K. "What Is a Neurotransmitter?" *about.com*

9. Purves, D., Augustine, G., Fitzpatrick, D., Katz, L., LaMantia, A., McNamara, J. and Williams, , S, Editors. *Neuroscience.* 2nd edition.2001, Sinauer Associates.

10. Baker K, Baldeweg T, Sivagnanasundaram S, Scambler P, Skuse D. "Met modifies mismatch negativity and cognitive function in 22q11 deletion syndrome." *Biol Psychiatry.* Jul 1,2005 58(1):23-31.

4
Traits

1. Menon, V. "Imaging study reveals differences in brain function for children with math anxiety" *Stanford Medicine News Center,* Mar 21, 2013.

2. Shumyatsky GP, Malleret G, Shin RM, Takizawa S, Tully K, Tsvetkov E, Zakharenko SS, Joseph J, Vronskaya S, Yin D, Schubart UK, Kandel ER and Bolshakov VY "Stathmin, A gene enriched in the amygdala, controls both learned and innate fear." *Cell.* 123:697–709. 2005

3. Audero, E., Mlinar, B., Baccini, G., Skachokova, Z., Corradetti,R. and Gross,C. "Suppression of Serotonin Neuron Firing Increases Aggression in Mice" *The Journal of Neuroscience,* 15 May 2013, 33(20).

4. Brocke B, Lesch KP, Armbruster D, Moser DA, Müller A, Strobel A, Kirschbaum C. "Stathmin, a gene regulating neural plasticity, affects fear and anxiety processing in humans." *Am J Med Genet B Neuropsychiatr Genet.* Jan. 5, 2010: 243-251.

5. Heinz A, Smolka MN, Braus DF, Wrase J, Beck A, Flor H, Mann K, Schumann G, Büchel C, Hariri AR, Weinberger DR. "Serotonin transporter genotype (5-HTTLPR): effects of neutral and undefined conditions on amygdala activation." *Biol Psychiatry.* Apr 15, 2007;61(8):1011-4.

6. Luoni, A., Hulsken, S., Cazzaniga,G., Racagni, G., Homberg, J. and Riva, M., "Behavioural and neuroplastic properties of chronic lurasidone treatment in serotonin transporter knockout rats." (1 July 2013) *International Journal of Neuropsychopharmacology.*

7. Lozier, L.,, Cardinale,E, VanMeter, J., and Marsh, A. "Mediation of the Relationship Between Callous-Unemotional Traits and Proactive Aggression by Amygdala Response to Fear Among Children With Conduct Problems." *JAMA Psychiatry.*_2014;71(6):627-636.

8. Scott, A., Bortolato, M., Chen, K. and Shiha, J. "Novel monoamine oxidase A knock out mice with human-like spontaneous mutation" *Neuroreport.* May 7, 2008; 19(7): 739–743.

9. Gallardo-Pujol, Andrés-Pueyo, Maydeu-Olivares, "MAO-A genotype, social exclusion and aggression: an experimental test of a gene-environment interaction." *Genes Brain Behavior.* Feb. 12, 2013, (1):140-5.

10. Reuter, M., Weber, B., Fiebach, C.J., Elger, C. and Montag, C. "The biological basis of anger: Associations with the gene coding for DARPP-32 (PPP1R1B) and with amygdala volume." *Behavioural Brain Research,* 202, 179-183. ( 2009 ).

11. de Almeidaa, R., Ferrarib,P.F., Parmigianic, S. and Miczekd, K., "Escalated aggressive behavior: Dopamine, Serotonin and GABA," *European Journal of Pharmacology,* Volume 526, Issues 1–3, 5 December 2005, Pages 51–64.

12. Miczek, K. "Monoamines, GABA, Glutamate, and Aggression, *Biology of Aggression, Oxford Scholarship Online,* May 2009.

13. Carré JM, McCormick CM, Hariri AR. "The social neuroendocrinology of human aggression," *Psychoneuroenocrinology.* Aug. 2011; 36(7):935-44.

14. Archer, J. "Testosterone and human aggression: an evaluation of the challenge hypothesis." *Neuroscience & Biobehavioral Reviews,* Vol 30, N3, 2006, 319–345.

15. Marcus DK, Zeigler-Hill V, Mercer SH, Norris AL "The psychology of spite and the measurement of spitefulness. " *Psychol Assess.* Jun26, 2014 (2):563-74.

16. de Quervain, , Fischbacher U, Treyer V, Schellhammer M, Schnyder U, Buck A, Fehr E. "The Neural Basis of Altruistic Punishment.", *Science* 27 August 2004: Vol. 305 no. 5688 pp. 1254-1258.

17. Morimoto, K., Miyatake, R., Nakamura, M., Watanabe, T., Hirao, T., and Suwaki, H. "Delusional Disorder: Molecular Genetic Evidence for Dopamine Psychosis." *Neuropsychopharmacology* (2002) 26 794-801.

18. Kahan, D. "What is Motivated Reasoning and How Does it Work." *Science and Religion Today,* May 4, 2011.

19. Freud S. *Inhibitions, Symptoms and Anxiety* (The Standard Edition) W. W. Norton & Company; 1926.

20. Russell, Bertrand, History of Western Philosophy. Simon & Schuster (1945).

21. Gyurak, A., Gross, J. and Etkin, A. "Explicit and Implicit Emotion Regulation: A Dual-Process Framework." *Cogn Emot.* 2011 Apr; 25(3): 400–412.

22. Shermer, M. "Logic-Tight Compartments." *Scientific American* January, 2013

23. Drew Westen, Pavel S. Blagov, Keith Harenski, Clint Kilts, and Stephan Hamann, "Neural Bases of Motivated Reasoning: An fMRI Study of Emotional Constraints on Partisan Political Judgment in the 2004 U.S. Presidential Election." *Journal of Cognitive Neuroscience,* November 2006, Vol. 18, No. 11, Pages 1947-1958

24. Inzlicht, M. and Gutsell, J. "Running on Empty- Neural Signals for Self-Control Failure" (2007) *Psychological Science,* Volume 18, Number 11,, pp 933-937.

25. Knafo A, Israel S, Darvasi A, Bachner-Melman R, Uzefovsky F, Cohen L, Feldman E, Lerer E, Laiba E, Raz Y, Nemanov L, Gritsenko I, Dina C, Agam G, Dean B, Bornstein G, and Ebstein RP. "Individual differences in allocation of funds in the dictator game associated with length of the arginine vasopressin 1a receptor RS3 promoter region and correlation between RS3 length and hippocampal mRNA." *Genes Brain Behav.* (3):266-75. 2008.

26. Bouchard, Jr, T., Lykken, D., McGue, M., Segal, N. and Tellegen, S., "Sources of Human Psychological Differences: The Minnesota Study of Twins Reared Apart." *Science,* Vol. 250, No. 4978, Oct. 12, 1990 p. 223-228.

27. Varki, A. and Brower, D. Denial: Self-Deception, False Beliefs, and the Origins of the Human Mind, *Twelve- Hachette Book Group,* June 4, 2013.

28. Bagemihl, B. Biological Exuberance: Animal Homosexuality and Natural Diversity, St. Martin's Press, 1999.

29. Goldstein, D. "Biological Basis of Sexual Orientation." *Stanford University News Service.* Mar, 3, 1995.

30. LeVay, S. "A difference in hypothalamic structure between heterosexual and homosexual men." *Science,* August 30, 1991, vol.253 no. 5023. pp. 1034–1037.

31. Savic, L. and Lindström P. "PET and MRI show differences in cerebral asymmetry and functional connectivity between homo- and heterosexual subjects." June 16, 2008. *Proc Natl Acad Sci USA* 105(27):9403-8.

32. Balthazart, J. *"Brain Development and Sexual Orientation" _Colloquium Series on The Developing Brain,* University of Liége. August 2012,

33. Malory, M. "Homosexuality & Choice: Are Gay People 'Born This Way?" *Scientific American* October 2013.

34. Liu,Y., Jiang,Y., Si,Y., Kim, J., Chen, Z-F. and Rao, Y. "Molecular regulation of sexual preference revealed by genetic studies of 5-HT in the brains of male mice." *Nature* 472, 95–99, 07 April 2011.

35. *http://exodusinternational.org,*

36. Bailey, R. "Is Heaven Populated Chiefly by the Souls of Embryos? Harvesting stem cells without tears." *reason.com* December 22, 2004

37. Boué A, Boué J, and Gropp A. "Cytogenetics of pregnancy wastage." *Adv. Hum. Genet.* 1985;14:1–57.

38. *BBC News, Europe,* 1 November 2013.

39. Ulrich Wagner andAndreas Zick "The relation of formal education to ethnic prejudice: Its reliability, validity and explanation.", *European Journal of Social Psychology* 22 Feb. 2006

40. Crick, Francis, The Astonishing Hypothesis p. 31, Charles Scribner's Sons, 1994

41. Reber, P. "What Is the Memory Capacity of the Human Brain?" *Scientific American,* June, 2010.

42. Adam, J., Calhoun, A. Tong, A., Pokala, N., Fitzpatrick,J. and Sjarpee, T. "Neural Mechanisms for Evaluating Environmental Variability in

*Caenorhabditis elegans.*" *Neuron* Volume 86, Issue 2, p428–441, 22 April 2015

43. Nabavi, S., Fox, R., Proulx, C., Lin, J., Tsien, R. and Malinow, R., "Engineering a memory with LTD and LTP." *Nature* 13294. June 1, 2014.

44. Ramirez, S., Liu, X., Lin, P-A., Suh, J., Pignatelli, M., Redondo, R., Ryan, T. and Tonegawa, S. "Creating a False Memory in the Hippocampus." *Science,*, Vol. 341.July 26, 2013

45. Rilling, J. and Berns, G. "Emory Brain Imaging Studies Reveal Biological Basis For Human Cooperation." *Neuron,* July 18, 2002.

46. Lutz, A., Brefczynski-Lewis, J., Johnstone, T., Davidson, R. "Regulation of the neural circuitry of emotion by compassion meditation: Effects of meditative expertise." *PLOS ONE, 3.* March 26, 2008.

47. Dunn, E. W., Aknin, L. B., and Norton. "Spending money on others promotes happiness." *Science, 319*, 2008, 1687–1688.

48. Piff, P., Stancatoa, D., Côtéb,S., Mendoza-Dentona, R. and Keltnera, D. "Higher social class predicts increased unethical behavior." *Proc Natl Acad Sci USA,* vol. 109 no. 11, Mar 13, 2012.

49. Saturn, S,R "Oxytocin Revisited-How I Learned to Stop Worrying and Love My Genes." *Cerebral Vortex,* Nov. 22, 2011

50. Kogana, A., Saslowb,L., Impetta,E., Oveisc, C., Keltnerd, D., and Saturne, S. "Thin-slicing study of the oxytocin receptor (OXTR) gene and the evaluation and expression of the prosocial disposition." *Proc Natl Acad Sci USA*, 2011 vol. 108 no. 48.

51. Keltner, D. "The Compassionate Instinct" *Greater Good,* Nov 05, 2014

52. Helminiak, D. The Human Core of Spirituality: Mind as Psyche and Spirit, State University of New York Press. 1996.

53. Fredrickson, B. "Why Choose Hope?" *Psychology Today,* March 23, 2009.

54. Bronowski, J. The Ascent of Man Little, Brown and Company, 1973.

55. Hamlin, J., Wynn, K. and Bloom, P. "Social evaluation by preverbal infants." *Nature* Vol 450, 22 November 2007.

56. Warneken, F. and Tomasello, M. "Altruistic helping in human infants and young chimpanzees.", *Science, 311,* 1301–1003. March 3, 2006

57. Calaprice A, ed. The Expanded Quotable Einstein. Princeton NJ: Princeton UniversityPress. *May 30, 2000.*

58. Schlegel, A., Kohler, P., Fogelson, S., Alexander, P., Konuthula, D. and Tse, P. "Network Structure and Dynamics of the Mental Workspace" *Proc Natl Acad Sci USA,* 110 (40) 16277-16282, (2013).

59. Matthias, G., German, B., Ranganath, C. "States of Curiosity Modulate Hippocampus-Dependent Learning via the Dopaminergic Circuit." *Journal Neuron,* 2014.08.060

60. Flaherty, AW "Frontotemporal and dopaminergic control of idea generation and creative drive." *J Comp Neurol* 493 (1): 147–53. (2005).

61. Danto, A. What Art Is Yale University Press, March 11, 2014.

62. Pike, A., Hoffmann, D., García-Diez, M., Pettitt, P., Alcolea, J., De Balbín, R., González-Sainz, C., de las Heras, C., Lasheras, J., Montes, R. and Zilhão, J. "U-Series Dating of Paleolithic Art in 11 Caves in Spain." *Science* Vol. 336 no. 6087 pp. 1409-1413. 15 June 2012:

63. Callaway, E. "*Homo erectus* made world's oldest doodle 500,000 years ago" *Nature News* December, 3, 2014.

64. Taylor, R., Spehar, B., Van Donkelaar, P. and Hagerhall, C. "Perceptual and Physiological Responses to Jackson Pollock's Fractals" *Frontiers in Human Neuroscience,* 2011; 5:60.

65. Bragg M. (1987). *Jackson Pollock.* TV documentary, South Bank Show, Independent Television.

66. Ramachandran, V.S. The Tell-Tale Brain: A Neuroscientist's Quest for What Makes Us Human, W. W. Norton & Company, *.2012.*

67. Ramachandran, V.S. and Hirstein, W. "The Science of Art, A Neurological Theory of Aesthetic Experience" *Journal of Consciousness Studies,* 6, No. 6-7, 1999, pp. 15–51

68. Vartanian O and Skov M, "Neural correlates of viewing paintings: evidence from a quantitative meta-analysis of functional magnetic resonance imaging data." *Brain Cogn.* June, 2014 87:52-6.

69. Smith, R. "The Well-Shaped Phrase as Art", *NY Times, Weekend Arts.* Nov 16, 2007.

70. Baumeister, R. F., & Leary, M. R. "The need to belong: Desire for interpersonal attachments as a fundamental human motivation." *Psychological Bulletin, 1995, 117,* 497–529.

71. Van Kesteren, R., Smit, A., Dirks, R., DeWith, N., Geraerts, W. and Joosse, J. "Evolution of the vasopressin/oxytocin superfamily: Characterization of a cDNA encoding a vasopressin-related precursor, preproconopressin, from mollucs *Lymnaea stagnates.*" *Proc. Nadl. Acad. Sci. USA* Vol. 89, pp. 4593-4597, May 1992.

72. De Dreu CK, Greer LL, Van Kleef GA, Shalvi S and Handgraaf MJ "Oxytocin promotes human ethnocentrism.", *Proc. Natl. Acad. Sci. USA.* **108** (4): 1262–1266, January 2011

73. Tomasello, T. Why We Cooperate *The MIT Press;* 2009

74. Crick, Francis; Koch, Christof (2005). «What is the function of the claustrum?". *Philosophical Transactions of the Royal Society B: Biological Sciences* 360 (1458): 1271–9.

75. "The Neurology of Consciousness, Crick's Last Stand." *The Economists* (July 28, 2005).

76. Stiefel, K.M. Merrifield, A. and Holcombe, A.O. "The claustrum's proposed role in consciousness is supported by the effect and target localization of *Salvia divinorum.*" *Frontiers in Integrative Neuroscience.* 26 February 2014.

5

Behavior

1. Dobzhansky, T. "Nothing in Biology Makes Sense Except in the Light of Evolution." 1973, *The American Biology Teacher* 35:125–29.

2. Drescher, S. *A History of Slavery and Antislavery, Cambridge University Press.* 2009

3. *http://www.globalslaveryindex.org*

4. Rejali, D. Torture and Democracy Princeton University Press (June 28, 2009).

5. *The United States Senate Select Committee on Intelligence,* July 17, 2002.

6. Nazaryan, A. "CIA Torture Report's Abu Zubaydah Surprised the Truth Came Out" December 16, 2104 *Newsweek.*

7. The Sentencing Project: "Report of The Sentencing Project to the United Nations Human Rights Committee Regarding Racial Disparities in the United States Criminal Justice System." *sentencingproject.org* August 2013

8. Kramer-Miller, B. "2 Prison Stocks That Look Good: Corrections Corp, Geo Group." *cheatsheet.com* April 7, 2014

9. The Editorial Board, *Vera Institute of Justice,* Nov. 26, 2014.

10. Benko, J. "The Radical Humaneness of Norway's Halden Prison" March 26, 2015. *NY Times.*

11. Erwin James, E. *The Guardian,* November 11, 2013.

12. Cooper, A., Durose, M. and Snyder, H. "Recidivism Of Prisoners Released In 30 States In 2005: Patterns From 2005 To 2010" *Bureau of Justice Statistics,* August 22, 2014.

13. Russell, B. Why Men Fight *The Century Co.* 1917.

14. "Greenhouse gas benchmark reached. Global carbon dioxide concentrations surpass 400 parts per million for the first month since measurements began." May 06, 2015 *National Oceanic and Atmospheric Administration.*

15. *AtmosNews, University Corporation for AtmosphericResearch,* May 15, 2013.

16. The Paleoclimatology Program, National Oceanic and Atmosphere Administration.

17. Shindell, D., Faluvegi, G., Bell, N. and Schmidt, G. "An emissions-based view of climate forcing by methane and tropospheric ozone." 2005 *Geophys. Res. Lett.*, 32.

18. Avery, J.A. "Quick Action Is Needed To Save The Long-Term Future" 11 February, 2015, *countercurrents.org*

19. Urban, M. "Accelerating extinction risk from climate change." *Science*: Vol. 348 no. 6234 pp. 571-573.May 1, 2015.

20. Jessica Aldred, J. "Caribbean coral reefs 'will be lost within 20 years' without protection." *The Guardian,* July 2, 2014.

21. Katharine Gammon "Half of Great Barrier Reef Lost in Past 3 Decades" *LiveScience* October 01, 2012

22. Ahmed, N. "US Navy predicts summer ice free Arctic by 2016" *The Guardian,* 9 Dec 2013.

23. *World Meteorological Organization.*

24. Henao, L. and Borenstein, S. "Glacial Melting in Antarctica Makes Continent The 'Ground Zero of Global Climate Change." *The HuffingtonPost* Feb 27, 2015.

25. Mooney, C. and Warrick, J., "Research casts alarming light on decline of West Antarctic glaciers", *The Washington Post,* December 4, 2014.

26. Greenbaum,J., Blankenship,D., Young,D., Richter, T., Roberts, J., Aitken, A., Legresy, B., Schroeder, D., Warner, R., van Ommen, T. and Siegert, M. "Ocean Access to a Cavity beneath Totten Glacier in East Antarctica" *Nature Geoscience,* 26 March, 2015.

27. 27. Hansen, J., Sato, M., Hearty, P., Ruedy, R., Kelley, M., Masson-Delmotte, V., Russell, G., Tselioudis, G., Cao, J., Rignot, E., Velicogna, I., Tormey, B., Donovan, B., Kandiano, E., von Schuckmann, K., Kharecha, P., Legrande, A. N., Bauer, M., and Lo, K.-W.: "Ice melt, sea level rise and superstorms: evidence from paleoclimate data, climate modeling, and modern observations that 2 °C global warming could be dangerous," *Atmos. Chem. Phys.*, 16, 3761-3812, doi:10.5194/acp-16-3761-2016, 2016.

28. Max, A. "New fresh water in Arctic could shift Gulf Stream" *phys.org* April 5, 2011.

29. *J. Hansen, M. Sato, P. Hearty, R. Ruedy, M. Kelley, V. Masson-Delmotte, G. Russell, G. Tselioudis, J. Cao6, E. Rignot, I. Velicogna, E. Kandiano, K. von Schuckmann, P. Kharecha, A. N. Legrande, M. Bauer, and K.-W. Lo. "Ice melt, sea level rise and superstorms: evidence from paleoclimate data, climate modeling, and modern observations that 2 ◦C global warming is highly dangerous." Atmos. Chem. Phys. Discuss.,* 15, 20059–20179, 2015

30. Gillis, J. "2015 Was Hottest Year in Recorded History, Scientists Say" *NY Times,* Jan. 20, 2016.

31. McCauley, D., Pinsky, M., Palumbi, S., Estes J., , Joyce, F. and Warner, R. "Marine defaunation: Animal loss in the global ocean." January 16,, 20*Science* Vol. 347 no. 6219.

32. "Turn Down Heat: Confronting the New Climate Normal." *2014 International Bank for Reconstruction and Development / The World Bank.*

33. Amrith, S. "The Bay of Bengal, in Peril From Climate Change." *NY Times,* Oct. 13, 2013.

34. Harris, G. "Coal Rush in India Could Tip Balance on Climate Change" *NY Times*, NOV. 17, 2014

35. Funk, M. "Windfall,The Booming Business of Global Warming" The Penguin Press, 2014.

36. Davenport, C. "Obama to Take Action to Slash Coal Pollution" *NY Times,* June 1, 2014.

37. Neuman, S. *NPR,* March 22, 2015.

38. Coral Davenport, "Nations Approve Landmark Climate Accord in Paris" *NY Times,* Dec 12, 2015.

39. "Survey finds 97% of climate science papers agree warming is man-made." *The Guardian*, 16 May 2013

40. Germain, T. "The Anti-Science Climate Denier Caucus: 113th Congress Edition" *Climate Progress.* June 26, 2013.

41. Myers, J. "Critics point to Inhofe's record." *Tulsa World,* November 30, 2002.

42. Corn, D. "Inhofe's Grand Climate Conspiracy Theory: It's All About Barbra Streisand" *Mother Jones,* Dec. 2, 2014.

43. Bump, P. "Jim Inhofe's snowball has disproven climate change once and for all." *The Washington Post*, February 26, 2015.

44. Diagnostic and Statistical Manual of Mental Disorders, 2013 *American Psychiatric Association.*

45. Jim Inhofe, *MapLight.*

46. Walsh, B. *"Why Climate-Change Denial Is So Powerful". Time Magazine.* Oct. 04, 2013,

47. "Koch Industries: Secretly Funding the Climate Denial Machine," *Greenpeace*

48. Karlin, M. "Koch Brothers Are Endangering the Planet by Funding Climate Change Denial." *Buzzflash at Truthout.* March 17, 2015

49. Brulle, R. "Institutionalizing delay: foundation funding and the creation of U.S. climate change counter-movement organizations." *Climatic Change,* Volume 122, Issue 4, 681-694, February 2014

50. Root, J. and Wiseman, T. "One-on-One Interview With Ted Cruz." *Texas Tribune.* March 24, 2015.

51. John Cook, J., Nuccitelli, D., Green, S., Richardson, M., Winkler, B., Painting, R., Way, R.,, Jacobs, P., and Skuce, A. "Quantifying the consensus on anthropogenic global warming in the scientific literature" 2013 *Environ. Res. Lett. 8* 024024.
and: "Ozone Depletion, Losing Earth's Protective Layer" *National Geographic Online*

52. Douglass, A., Newman, P. and Solomon, S. "The Antarctic ozone hole: An update". _*Physicstoday.* July, 2014, Volume 67, Issue 7.

53. "Reviving the Ocean Economy: The case for action - *2015* " *World Wildlife Fund,* April 22, 2015.

54. Dangerfield, W. "The Mystery of Easter Island" *smithsonian.com* March 31, 2007.

55. Ponting, C. A New Green History of the World: The Environment and the Collapse of Great Civilizations. Penguin Books. *2007,*

56. Diamond, J. Collapse: How Societies Choose to Fail or Succeed. *Penguin Books.* 2011.

57. *agroecology.org*

58. *Wikipedia;* List of environmental and conservation organizations in the United States.

59. The United States Census Bureau, *census.gov*

60. Ehrlich, P. The Population Bomb, Buccaneer Books, Re-published 1995.

61. World Hunger Education Service, http://www.worldhunger.org

62. Hardin, G. "Living on a Lifeboat" September, 1974, *Bioscience* 24 (10), 561-568.

63. World energy consumption - *Wikipedia.*

64. "International Energy Outlook 2014." *U.S. Energy Information Administration.* Sept 9, 2014

65. 2014 Half-year Report. *The World Wind Energy Association* (2014) WWEA. pp. 1–8.

66. Fountain, H. "Nuclear: Carbon Free, but Not Free of Unease" *NY Times,* Dec. 22, 2014.

67. Schwartz, J. "Natural Gas: Abundance of Supply and Debate" Dec. 22, 2014 *NY Times.*

68. Frank, A. "Is a Climate Diaster Inevitable?" *NY Times.* Jan. 17, 2015.

69. Wind turbine with the world's largest rotor goes into operation. *Siemens* October 8, 2012.

70. Gillis, J. "A Tricky Transition From Fossil Fuel: Denmark Aims for 100 Percent Renewable Energy" *NY Times.* Nov. 10, 2014,

71. Davenport, C. "Obama's Strategy on Climate Change, Part of Global Deal, Is Revealed." *NY Times.* March 31, 2015.

72. Frank, R. "The Rich Are Less Charitable Than the Middle Class: Study." *CNBC.com.* August 20, 2012

73. Stern, K. "Why the Rich Don't Give to Charity" *The Money Report; The Atlantic.* April 2013,

6
Constructs

1. Dunbar, R. The Human Story: A New History of Mankind's Evolution, Faber & Faber. *2005*

2. Kitcher, P. Life After Faith: The Case of Secular Humanism *The Terry Lecture Series.* Yale University Press. Oct 28, 2014.

3. Zuckerman, M., Silberman, J. and Hall, J. "The Relation Between Intelligence and Religiosity." August 6, 2013, *Personality and Social Psychology Review.*

4. McNall Burns, E., Ralph, P,. Lerner, R., and Meacham, S. World Civilizations, Seventh Edition, Vol. 1, 1986, W.W. Norton & Company.

5. Avery, J.S. "The Future of International Law." *countercurrents.org* April 18, 2015

6. *Wikipedia,* Kevin O'Lear

7. *RenewAmerica,* December 16, 2011

8. Lindstrom, L. "The Good, the Bad, and the Ugly in the Student-Loan Deal.." *The Chronicle of Higher Education.* August 8, 2013

9. Goleman, D. "Rich People Just Care Less" Oct, 5, 2013, *Opinion Pages, NY Times.*

10. *10.* Allison, B. and Harkin, S."Fixed Fortunes: Biggest corporate political interests spend billions, get trillions." *Sunlight Foundation.* Nov. 17, 2014.

11. Demko, P. "How healthcare's Washington lobbying machine gets the job done." *Modern Healthcare,* October 4, 2014.

12. Evo Morales: "Ten commandments against capitalism, for life and humanity." October 9, 2008, Solstice celebration in Lake Titicaca, Bolivia.

13. Planas, R. 'World's Poorest President' Rages Against The Necktie, Calling It A 'Useless Rag' " *The Huffington Post.* May 21, 2014,

14. Thompson, D. "How Low Are U.S. Taxes Compared to Other Countries?" *The Atlantic.* Jan 14, 2013

15. "World Happiness Report, *2013"* Edited by Helliwell, J., Layard, R and Sachs, J. *UN General Assembly.*

16. George McGovern The Essential America. Simon & Schuster, 2004.

17. Gilens, M. and Page, B. "Testing Theories of American Politics: Elites, Interest Groups, and Average Citizens." *Perspectives on Politics* / Volume 12 / Issue 03, Sept 2014, pp 564-581.

18. Smith, E. and Kuntz, P. "CEO Pay 1,795-to-1 Multiple of Wages Skirts U.S. Law" *Bloomberg Business.* April 29, 2013

19. Wearden, G. "Oxfam: 85 richest people as wealthy as poorest half of the world" *The Guardian.* January 20, 2014

20. "Outlook on the Global Agenda 2014" *2013 World Economic Forum.*

21. Saez, E. "Striking it Richer: The Evolution of Top Incomes in the United States." September 3, 2013 *Pathways Magazine,* Stanford Center for the Study of Poverty and Inequality.

22. Lepore, J. "The Crooked and the Dead. Does the Constitution protect corruption?" Aug. 25, 2014 *The New Yorker.*

23. Thompson, D. "How Low Are U.S. Taxes Compared to Other Countries?" *The Atlantic.* January 14, 2013,

24. *Daytona Beach Morning Journal,* May 8, 1971.

25. Hatemi, P., Gillespie, N., Eaves, L., Maher, B., Webb, B., Heath, A., Sarah E. Medland, S., Smyth,D., Beeby, H., Gordon, S., Grant W. Montgomery, G. and Byrne, E. "A Genone-Wide Analysis of Liberal and Conservative Political Attitudes." January 2011 *The Journal of Politics,* Vol. 73, No. 1.

26. Safra, E. and Dan M. Kahan, D. "Climate Science Communication and the Measurement Problem." *Advances Pol. Psych,* June 25, 2014.

27. Gardner,H. Changing Minds: The Art and Science of Changing Our Own and Other People's Minds. *Harvard Business School Press.* 2004,

28. Perry, S. *Psychology Today,* May 15, 2011.

29. 29, C.P. Snow The Two Cultures and the Scientific Revolution - The Rede Lecture, 1959 Cambridge University Press, 1961.

30. Harris, R. "U.S. Science Suffering From Booms And Busts In Funding." *NPR* Sept. 9, 2014

31. "Dr. No Money: The Broken Science Funding System" *Scientific American* April 11, 2011.

32. Deng, B. "Congress Is Terrible at Science—and This Should Make Us Worried." *slate.com* May 7, 2014

33. Editorial, Opinion Pages. "Trifling With the Arts and Humanities" August 20, 2013, *NY Times.*

34. National Endowment for the Arts Fact Sheet Spring 2013 *nasaa-arts.org*

35. Zakaria, F. In Defense of a Liberal Education, W.W. Norton & Company, March 30, 2015

Quotations are from a variety of sources including author's words as published and www.brainyquote.com, www.goodreads.com and www.searchquotes.com